Comprehensive Quality by Design in Pharmaceutical Development and Manufacture

Topical Conference at the 2011 AIChE Annual Meeting

Minneapolis, Minnesota, USA
16-21 October 2011

ISBN: 978-1-61839-584-9

Printed from e-media with permission by:

Curran Associates, Inc.
57 Morehouse Lane
Red Hook, NY 12571

Some format issues inherent in the e-media version may also appear in this print version.

Printed by Curran Associates, Inc. (2012)

AIChE
3 Park Avenue
New York, NY 10016-5991

Phone: (203) 702-7660
Fax: (203) 775-5177

www.aiche.org

Additional copies of this publication are available from:

Curran Associates, Inc.
57 Morehouse Lane
Red Hook, NY 12571 USA
Phone: 845-758-0400
Fax: 845-758-2634
Email: curran@proceedings.com
Web: www.proceedings.com

TABLE OF CONTENTS

(48a) Quality by Design (QbD) Opportunities, Challenges and Future Direction

Moheb Nasr, DHHS/FDA/Cder/OPS/Ondqa, FDA, Silver Spring, MD

QbD is a systematic approach to pharmaceutical development, manufacturing and product lifecycle management. QbD provides new opportunities to utilize state of the art science, quality risk management and modern engineering tools. This presentation will discuss current regulatory guidelines and implementation progress. In addition, Dr. Nasr will outline some of the challenges and provide suggestions to move forward. The regulatory role need to be better defined to encourage innovation, facilitate drug development and provide high assurance of quality.

See more of this Session: QbD Plenary
See more of this Group/Topical: Topical I: Comprehensive Quality by Design in Pharmaceutical Development and Manufacture

(48b) QbD - It's Not Just for Breakfast Anymore

Paul C. Collins, Chemical Product Research and Development, Eli Lilly and Co, Indianapolis, IN

We are approaching the ten-year mark for the Quality by Design initiative in the pharmaceutical industry. Much has been accomplished and demonstrated, particularly with regard to enhanced understanding of processes and better definition of processing parameters such as CPPs. Design space has become an often utilized term when process scientists talk about their projects and the concept has led to increased use of modeling approaches in late stage development. Despite the interest generated by the initiative and the progress made by it's participants, the true power of QbD may yet to be unleashed. Human-centered design of pharmaceuticals, with a more intimate tie to patient need and patient performance, should be the next step in how we more forward the QbD initiative, and represents a new way for engineers to influence drug design.

See more of this Session: QbD Plenary
See more of this Group/Topical: Topical I: Comprehensive Quality by Design in Pharmaceutical Development and Manufacture

(48c) Challenges and Opportunities for Process Systems Engineering Approaches to QbD and Design Space

G. V. Reklaitis, Pharmaceutical Technology & Education Center, Purdue University, W Lafayette, IN

The key components of the QbD concept, namely, the Design Space and PAT, do draw extensively for their realization on tools from the domain of process systems engineering methodology. Both the challenges and the opportunities in adapting core PSE methodology, such as process modeling and simulation, process dynamics and control, monitoring and fault diagnosis, quantification of process uncertainty and risk, and product and process optimization, to the development and manufacture of pharmaceutical products are significant. In this presentation we will review the progress made in the application of pse methods to this domain and outline some of the important developments that remain to be realized.

See more of this Session: QbD Plenary
See more of this Group/Topical: Topical I: Comprehensive Quality by Design in Pharmaceutical Development and Manufacture

Practical Applications of QbD for a Parenteral Drug Product

Nancy J. Harper, Venkat Koganti and Gautam R. Ranade, Pharmaceutical Research and Development, Pfizer, Groton, CT

This presentation addresses the application of Quality by Design principles to two common unit operations encountered in the manufacture of parenteral products. Examples will be described for 1) the application of mixing models to facilitate the scale-up of the API dissolution step, and 2) an approach for establishing a sterilization design space based on chemical kinetics and sterilization theory.

API Dissolution During Solution Compounding

In this example, mixing conditions to achieve complete dissolution of an API are dependent on several process variables, including API particle size, temperature, agitation rate, agitation time, and tank/agitator configuration. The fluid dynamics of the mixing vessel are scale-dependent, but can be modeled to enable predicted mixing conditions on scale-up. The approach taken in this example was to develop a lab-scale mixing vessel that was geometrically similar to the production vessel. The key aspect ratios of the production equipment (impeller diameter/tank diameter, liquid height/tank diameter, distance between impeller/tank diameter) were reproduced in a lab-scale mixer. Based on fluid dynamic principles, maintaining a similar power per unit volume (P/V) during mixing at either scale will result in similar solid-liquid mass transfer coefficient and therefore similar mixing times to achieve dissolution. Determinations of P/V were made using computational fluid dynamic (CFD) modeling of the mixing systems. Dissolution studies were performed in the lab-scale mixer to evaluate the combined influences of agitation rate and batch temperature on dissolution time, using worst-case API (a batch with particle size parameters near the maximum control limits). Using CFD modeling to determine agitation rates that will achieve the same P/V in the commercial tank, along with a constant scaling factor statistically determined during development of the model for this tank geometry, predicted agitation rates were determined that would achieve complete dissolution within the same time interval as observed in lab-scale. The suitability of the predicted agitation rates was determined for one full-scale batch, intentionally manufactured at the lower limit of the temperature range, representing worst-case conditions for dissolution. Successful dissolution within the target time interval served to confirm the model predictions and verify the minimum agitation rate and time for commercial production.

Terminal Sterilization

The molecule for this example is heat sensitive, and when exposed to pharmacopoeial (overkill) sterilization conditions (15 minutes at 121 °C), the extent of degradation exceeds an acceptable impurity level. Therefore a temperature below 121°C was evaluated for the sterilization cycle. Both microbiological lethality and degradant formation are directly dependent on cumulative thermal exposure, and therefore sterilization conditions are well suited for the development of a design space. To establish the design space, sterilization temperature and exposure time were explored in terms of conditions that simultaneously provide 1) a minimum $\log_{10}$ reduction of 8 (assuring sterility), and 2) maximum degradant formation according to the specified limit (assuring appropriate purity).

This approach to the establishment of a design space is unique in that it allows the use of first principles (chemical kinetics and sterilization theory) to define acceptable boundaries for process parameters. While some experimentation is required (i.e., determination of degradation rate constants and sterilization D-value), these studies are scale-independent and can be conducted at the lab bench. The experimentally determined constants are incorporated into theoretical equations that are used to define the boundaries of the design space.

The degradation rate constants were determined in development studies by traditional chemical kinetic experimentation utilizing drug product exposed to elevated temperatures in the range of sterilization conditions (above 100°C). Rate constants were determined at worst-case conditions of pH and headspace oxygen content based on targeted in-process control limits, thus representing worst-case stability conditions within the formulation design space. The Arrhenius equation was then used to define a time/temperature profile that results in a pre-defined maximum limit of degradation. For sterilization requirements, the D-value of a thermally resistant spore-former (*Geobacillus stearothermophilus*) was determined in the drug product solution. Based on sterilization theory, a time/temperature profile was derived that would result in an 8 log reduction of microorganisms. Superposition of these two time/temperature profiles then defines the design space, shown in Figure 1.

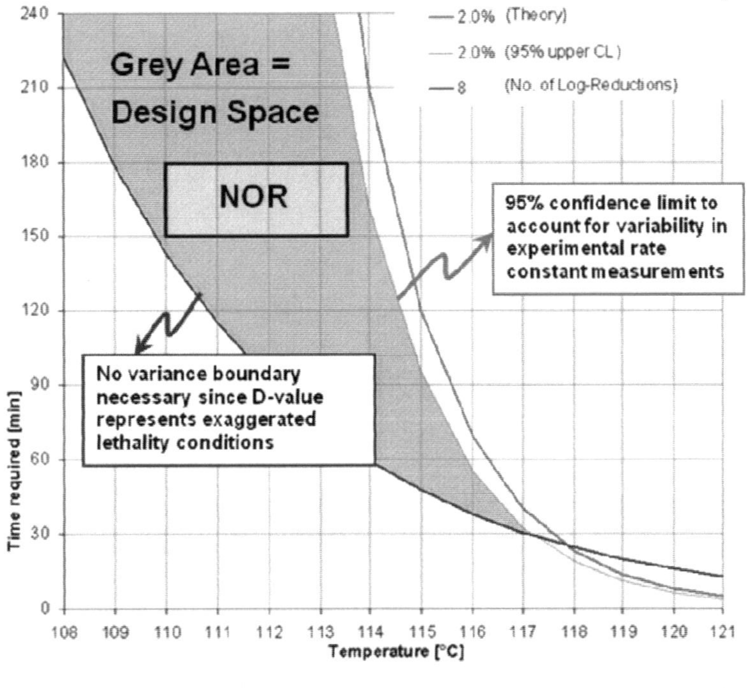

Figure 1. Moist Heat Sterilization Design Space and Normal Operating Ranges

The design space definition includes the 95% confidence limits for degradation. Statistical variance considerations for the lethality input (log-reduction) were not applied since the lethality experiments (D value generation) were performed with highly heat-resistance spores (*Geobacillus stearothermophilus*) to reflect exaggerated conditions.

Acknowledgements

Mixing Model: Fred Carroll, Rich Ferraina, Mark Berry, Yang Liu, Ken Norris, Robert Leasure, Jeff Gaudio (Pfizer); Rick Falk, Yogesh Waghmare (Bend Research)

Sterilization: EFPIA Mock P2 Injectables group: Heinz Wiederkehr (Roche), Thomas Backensfeld (Bayer Schering), Ritva Haikala (Orion Pharma), Christian DeMuynck (Nycomed), Brian Withers (Abbott)

(68b) Rational Design and Scale-up of Batch Lubrication Processes for Immediate Release Formulations

Joseph Kushner IV, Daniel O. Blackwood, Bruce C. MacDonald, Mark Polizzi, Francis M. Moore and Bruno C. Hancock, Pharmaceutical Development, Pfizer Worldwide Research and Development, Groton, CT

Purpose: To provide stepwise guidance for lubrication of pharmaceutical powders through the development of a unifying model which captures lubrication effects during blending and compression.

Methods: Literature models of powder properties following lubrication blending over a range of scales and conditions were reviewed. Additionally, a new model was developed by lubricating pharmaceutical powders in a tablet press feed frame. Lubricated powders were compressed on a development-scale rotary tablet press (Kilian T100) with varying throughput and feed-frame paddle rotation speeds. The resulting tablets were measured for tensile strength and a feed frame lubrication sensitivity model was developed. The two models were then combined to create an integrated lubrication understanding framework.

Results: Lubrication process design and scale-up guidelines were developed that addressed the selection of the appropriate amount of lubricant, the impact of equipment design (scale & principle of operation), and the need to select processing conditions that will result in the consistent manufacture of product possessing the desired extent of lubrication. To provide quantitative guidance to this end, a new unified model was developed which captured the effect of lubrication on pharmaceutical powders during blending and compression.

Conclusions: The proposed stepwise approach outlines the current state-of-the-art for conventional tablet blend lubrication processes, and embodies the principles of Quality-by-Design and risk-based drug product development. As such it is recommended as the starting point for scientists undertaking the development of immediate release tablet formulations.

See more of this Session: Application of Quality by Design In Drug Product Process Development I
See more of this Group/Topical: Topical I: Comprehensive Quality by Design in Pharmaceutical Development and Manufacture

(68c) Automation and Control of Drug-On-Demand Technology

Laura Hirshfield, Arun Giridhar, Gintaras V. Reklaitis, Venkat Venkatasubramanian and Michael T. Harris, Chemical Engineering, Purdue University, West Lafayette, IN

After the implementation of initiatives such as PAT and QbD, there has been a drive towards efficient manufacturing in the pharmaceutical industry. The concept of "mini-manufacturing" of drugs, as opposed to mass production of drugs, has shown promise in this area. We describe one such "mini-manufacturing" process, "drug-on-demand," which uses drop-on-demand printhead technology to deposit active pharmaceutical ingredients (API) onto edible substrates. This method also allows for the layering of different APIs and the creation of personalized dosage forms, in which the amount of API can be varied depending on the patient. In this work, we focus on the automation and control strategy for the drug-on-demand pilot facility done as part of the Engineering Research Center for Structured Organic Particulate Systems. A proper control strategy is necessary so that the system can run at optimal setpoints and adjust in real-time, allowing for an efficient manufacturing process and a precise on-spec drug product. The control strategy for our "drug-on-demand" setup allows us to execute an automated, optimized, and controlled print cycle while closely monitoring drop size, drug morphology, and drop deposition pattern.We summarize our achievements and current research in this area.

See more of this Session: Application of Quality by Design In Drug Product Process Development I
See more of this Group/Topical: Topical I: Comprehensive Quality by Design in Pharmaceutical Development and Manufacture

Dynamic simulation of the dissolution of pharmaceutical tablet integrated with continuous manufacturing process

Presenter: Lukasz Mioduszewski

Research Participants: Fani Boukouvala, Marianthi Ierapetritou, Rohit Ramachandran

ENGINEERING RESEARCH CENTER FOR
STRUCTURED ORGANIC PARTICULATE SYSTEMS
RUTGERS UNIVERSITY
PURDUE UNIVERSITY
NEW JERSEY INSTITUTE OF TECHNOLOGY
UNIVERSITY OF PUERTO RICO AT MAYAGÜEZ

Scientific Goals

- Develop a dynamic model of dissolution of pharmaceutical tablets as part of a dynamic flowsheet simulation of a continuous tablet manufacturing process
- Correlate operating and design parameters of upstream processes to properties of tablet dissolution through the integration of processing units
- Develop a model for tablet dissolution using Fick's Second Law with a non-constant diffusion coefficient

Introduction

Flowsheet simulation:

❑ Flowsheet simulation is used for **process design**, dynamic optimization and flexibility analysis of complex large-scale integrated systems.
❑ Application of simulators for solid processes is still very primitive compared to state of the art for fluid processes.
❑ Through dynamic flowsheet simulation of an integrated tablet manufacturing line, operating conditions and design of the system can be linked to properties of final product
❑ The effects of possible perturbations during operation can be and the necessary control scheme to alleviate them can be identified

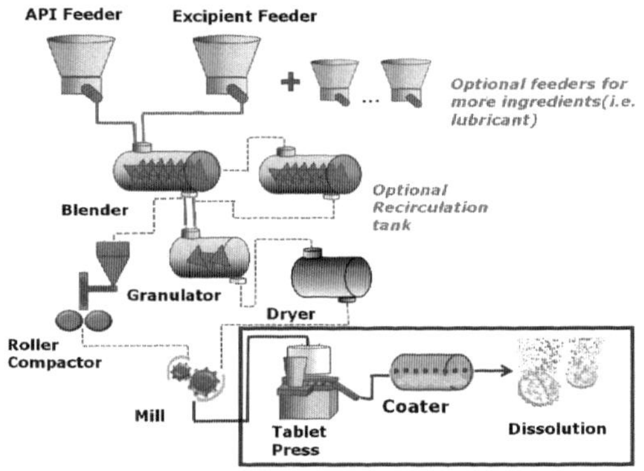

Tablet Dissolution:

A tablet is ingested and passed through the esophagus to the stomach, which is an aqueous environment. This is the first place where the tablet may dissolve.
The rate of dissolution is a key target for controlling the duration of a drug's effect.

Methods:
Representation of tablet:

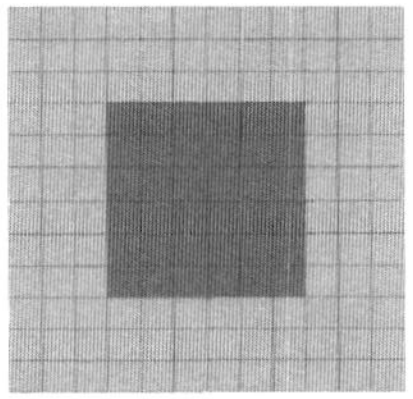

We initialized the void fractions with the equation for porosity. We assigned the center of the tablet a value of void fraction 20% larger than that for the outer tablet volume

Governing equations dissolution:

Fick's Second Law

$$\frac{\partial c_i}{\partial t} = -\nabla \cdot (-D_i \nabla c_i) + s_i \quad i = A,B$$

where

Source term

$$s_i(\mathbf{x}) = \begin{cases} k_i(c_i^{sat} - c_i(\mathbf{x})) & \forall \mathbf{x}: \phi_i(\mathbf{x}) > 0 \\ 0 & \text{otherwise} \end{cases}$$

Diffusion coefficient

$$D_i = D_{i0}\left(1 - \sum_{j=1}^{n} \phi_j\right)^{\alpha_i}$$

Evolution equation

$$\frac{\partial \phi_i}{\partial t} = -\frac{s_i}{\rho_i}$$

Where:
C_i is the concentration of component i
D_i is the diffusion coefficient
s_i is the source term
A,B are components (API and Excipient)
k_i is the dissolution rate constant
φ is the phase volume fraction (porosity)
α is diffusivity exponent
ρ is density

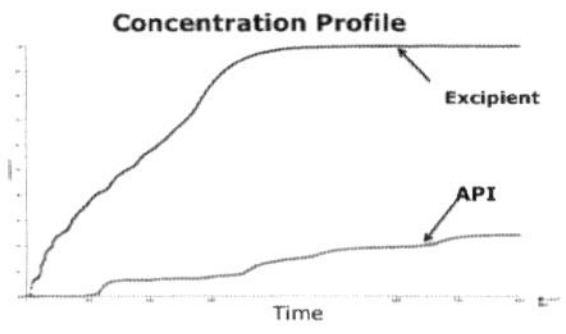

Integration of tablet press and dissolution

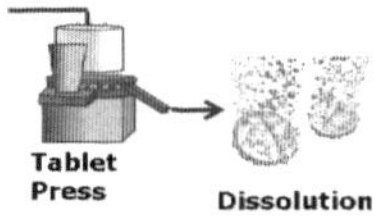

Correlation between compression force (F) of tablet press to porosity of tablet

φ= 0.5921*EXP(-0.0002*F)

Developed from empirical analysis (Nokhodchi, A. Ford, J. Rowe, P. Rubinstein, M.)

Correlation between mean particle size and Diffusion coefficient

$D_{i0} = kT / 6\pi\eta R_H$

Boltzmann constant, Temperature, Viscosity, Particle radius

Brownian Motion

Results:
Graphical Profiles of Results:

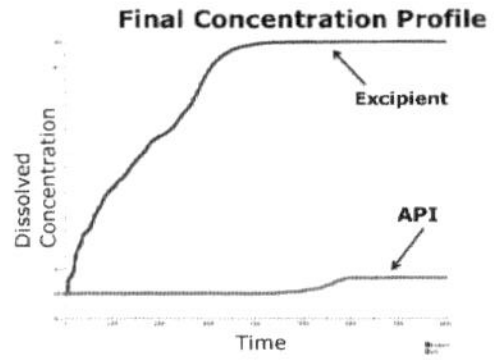

The concentration profiles for both API and Excipient increase until entire quantity is dissolved

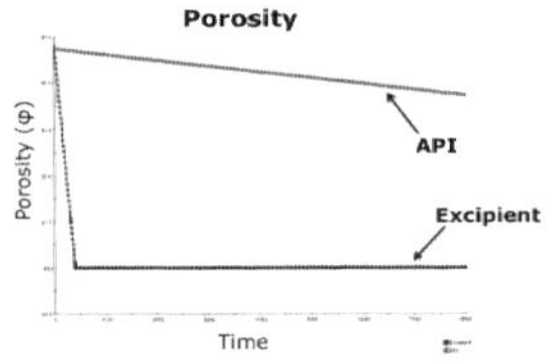

The porosity tends to zero for both API and Excipient. Profile for API tends to zero slowly because API volume fraction is initially higher than that of Excipient. Also, the API is located in the middle of the tablet, therefore its dissolution takes longer

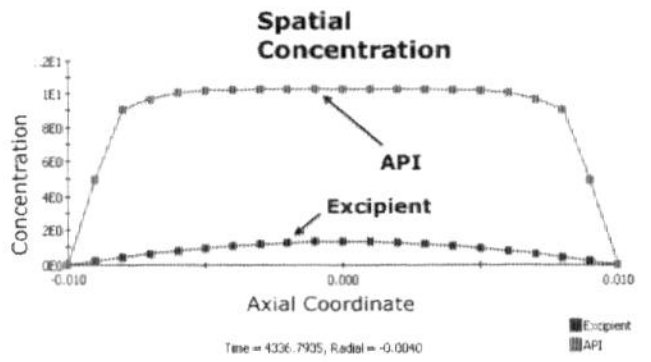

The spatial concentration profile of the tablet is shown. The concentration is maximum in the middle of the tablet and tends to zero when approaching the edges of the tablet

Conclusions

1. Dissolution model is the last step in the flowsheet simulation and determines many important properties of the final product
2. Tablet press and dissolution models were connected using Brownian motion and empirical data available for hydroxypropylmethylcellulose in regard to the compaction properties and void fractions
3. Flowsheet simulation of integrated system can help identify optimum design, operating conditions and sequence of tablet manufacturing process in order to produce high quality product with consistency

Future Work

1. Conducting experiments with the actual tablet press and obtaining real data for a specific pharmaceutical component
2. Modifying model parameters for a specific drug
3. Modifying the empirical equations in the model to customize the flowsheet for particular compounds

(68e) Innovative Techniques for Planning Experiments Across Mulitple Unit Operations

Kelly Canter, Pharmaceutical Sciences, Pfizer Inc., St. Paul, MN and Leonard Perry, University of San Diego, San Diego, CA

Many well-established analytical tools and methodologies exist for developing and characterizing a product for manufacture. However, the application of these tools all too often can fall short of their objective or full potential to gain process understanding – even in the hands of a well trained chemometrician or statistician. Another essential and uncommon taught skill is in the planning of experiments over the lifecycle of a products development and subsequent manufacture. The selection of the proper experimental approach and planning activities that precede the actual experimentation are critical to successfully establish process understanding. The application of proper risk management tools along with established experimental planning principles aide significantly in effectively establishing a product's experimental development plan. There are well documented sources that aid in the planning and selection of designed experiments to achieve a desired outcome, especially when looking at a single unit operation at a time. Yet minimal attention is given to the planning and sequential approach required for experimentation across multiple units operations. This paper highlights a set of tools and guidelines for assessing technical issues and establishing an appropriate experimental plan when looking across a products manufacturing process. A pharmaceutical case study involving the experimental strategy taken to develop the design space for a Quality-by-Design filing is provided, and the trade-offs between resource limitations and statistical consequences is included.

See more of this Session: Application of Quality by Design In Drug Product Process Development I
See more of this Group/Topical: Topical I: Comprehensive Quality by Design in Pharmaceutical Development and Manufacture

(68f) Real-Time Process Management and Knowledge Management for Continuous Pharmaceutical Manufacturing

Girish Joglekar, **Arun Giridhar**, Gintaras V. Reklaitis and Venkat Venkatasubramanian, Chemical Engineering, Purdue University, West Lafayette, IN

The manufacture of pharmaceutical drug products has benefited significantly from a QbD approach, including the enabling of continuous production over batch processing. Developing and improving processes for continuous manufacture requires the combined use of process data with process and formulation knowledge. Process data can be generated across multiple unitops simultaneously, and correcting deviations in one critical process parameter or critical quality attribute may require manipulations of a totally different process unit. Such integrated control must be done in a manner that is both easy to scale to different sizes and easy to transfer to different materials, requiring an integrated approach in gathering and managing process knowledge and using it flexibly for new processes.

In this work, we present the results and case studies from the Engineering Research Center for Structured Organic Particulate Systems (ERC-SOPS), in developing a real-time process management system for continuous pharmaceutical manufacture, that makes use of an ontological informatics framework to manage process knowledge of multiple types, including material properties (such as stress-density relationships), process parameters (such as process setpoints), equipment parameters (such as die and punch sizes in a tablet press), and process control knowledge (mathematical models for PID and model predictive control, process signatures for fault diagnosis, mathematical models to plan future setpoint changes, etc). The models taken together operate on a wide range of time scales, and the knowledge management framework is designed to be flexible for all uses. Further, we show how the system architecture allows for the use of various control systems and modeling programs in a flexible manner. Finally, we present case studies in tablet manufacturing and drop-on-demand systems, and summarize our current results.

See more of this Session: Application of Quality by Design In Drug Product Process Development I
See more of this Group/Topical: Topical I: Comprehensive Quality by Design in Pharmaceutical Development and Manufacture

(138a) Linkage of Critical Quality Attributes Across a Multi-Unit Operation Process Train Using a High-Shear Wet Granulation Design of Experiment

Diana S. Hou, Julianne Farabaugh, Gina Thompson and David Lavrich, Merck Manufacturing Division, Merck & Co., West Point, PA

As the momentum for implementation of Quality by Design progresses in the pharmaceutical industry, more tools are being implemented to create better understanding of process parameters and their impact on critical quality attributes. Utilizing models and statistics have become a key instrument in creating a robust design space and expansive knowledge space of the process trains. While doing one variable at a time experiments can be used to understand the ranges of parameters, it does not exemplify understanding of the interactions across an entire unit operation. Thus, there is a need to utilize design of experiments and statistical modeling within a unit operation to justify ranges on key processing parameters. However, across multiple unit operations, the initial parameters studied can be confounded by the effects of the other unit operations. An example of this is in a high-shear wet granulation (HSWG) process train, where multiple unit operations are needed in order to process the material into the final product, where the critical quality attributes can be tested. Here we present a case study on a design of experiments on a HSWG process train and how the initial outputs from the experiment are linked to the critical quality attributes without being confounded by the unit operations following the granulation. Due to the multi-unit operations needed in the HSWG train, we present a way to map the processing parameters of the granulation step to the critical quality attributes by using the granule characteristics as the linkage. By understanding interactions of the processing parameters on the granule characteristics and the variation of the granule characteristics on the critical quality attributes, the confounding effects across the other unit operations can be eliminated.

See more of this Session: Application of Quality by Design In Drug Product Process Development II
See more of this Group/Topical: Topical I: Comprehensive Quality by Design in Pharmaceutical Development and Manufacture

(138b) An Integrated Quality by Design (QbD) Approach towards Prediction of Protein Degradation In Heated Mixing Vessels by Computational Fluid Dynamics

Marco Iannuccelli[1], **Siegfried Adam**[1], Daniele Suzzi[1] and Johannes G. Khinast[2], (1)Research Center Pharmaceutical Engineering (RCPE), Graz, Austria, (2)Institute for Process and Particle Engineering, Graz University of Technology, Graz, Austria

Combined heating and mixing of protein solutions is a process often found in the pharmaceutical industry during the manufacture of biopharmaceutics. In the case of proteins or enzymes, the local anisotropy of temperature and the exposure time to such temperature may have a major effect on the degradation of the API(s) in the processed solution or emulsion and consequently, the quality of the biopharmaceutical medicinal product could be jeopardized. Hence, in order to assure patient safety and efficacy of the product and to avoid out-of-specification results leading to rejection of large amounts of costly compounds, optimal combinations of mixing operational parameters have to be evaluated.

In our study, a combined QbD and computer simulation approach is employed to (1) define a representative critical quality attribute of a biopharmaceutical manufacturing process (i.e., thermal degradation of the active pharmaceutical ingredient), (2) identify potentially critical input factors that may affect this critical quality attribute and (3) define and perform activities for computer simulation based process characterization. Data are then used to map out a knowledge space, providing parameters to define a design space as its subset to set up an appropriate control strategy.

In our work rotational speed and solution viscosity have been considered as potentially critical input factors leading to inhomogeneous temperature distribution resulting in increased degradation of the mixture components. Simulations have been performed according to statistically designed experiments. The degradation laws, which apply to particular enzymes and proteins, are well known from literature and have been directly integrated in the simulation model. The effects have been represented by means of a degradation index.

Results show that some region of the processed solution could present stagnation points or recirculation zones where energy cannot be exchanged by convective motions. On the contrary, in other regions higher velocity gradients, hence stronger convective motions can be detected. These fields are the main causes of the inhomogeneous temperature distribution (hot and cold spots) resulting in increased degradation of the mixture components. It is shown that the combination of lower rotational speed and higher viscosity provides the highest degradation index for the biological solution.

Data and knowledge generated in our study can be used as a generic approach to design efficient heated mixing processes for different API-excipients mixtures providing optimal performance and quality output. Moreover, the developed correlations between degradation of mixture components as one major critical quality attribute of biopharmaceutical manufacturing processes and critical process parameters can result in better quality assurance and validation strategies, taking into account more efficient in-process-controls and continued process verification throughout the life cycle of the (bio-)pharmaceutical product.

See more of this Session: Application of Quality by Design In Drug Product Process Development II
See more of this Group/Topical: Topical I: Comprehensive Quality by Design in Pharmaceutical Development and Manufacture

(138c) Applications (and Challenges) of Utilizing QbD, DOE and Other Process Tools In the Transfer of a Sterile Lyophilized Product From One Site to Another - Practical Perspectives

Daniel Kim, Manish Sharma, David Unger and **Stelios Tsinontides**, Worldwide Process Development, Cephalon, Inc., West Chester, PA

The FDA guidance on QbD has been in place for few years now, and one would anticipate that several manufacturers of medicinal products are applying it, or at least concepts from it, in the development and scale-up of manufacturing processes. This presentation will describe how concepts of QbD and DOE were successfully used in the scale-up of a lyophilized product to manufacturing. The presentation will also identify areas where these concepts of QbD and DOE encountered challenges in their application to yield anticipated results. The presentation will highlight the importance of utilizing these tools with a systematic approach and having always in mind that no change is a simple change without proof. Scale-up and transfer of manufacturing processes invlove the interplay of meriad factors, thus they need to be based on sound scientific and engineering principles and careful case by case risk analysis. Often enough small changes in processing could result to un-anticipated results, and such changes might not have been fully explored for one or another reason in a careful QbD or DOE studies at the laboratory scale. Robust manudacturing processes require application of science, engineering, and risk analysis tools on a continuous basis, from development all the way to and during commercial manufacturing.

See more of this Session: Application of Quality by Design In Drug Product Process Development II
See more of this Group/Topical: Topical I: Comprehensive Quality by Design in Pharmaceutical Development and Manufacture

(138d) Dissolution of Disintegrating Solid Dosage Forms In a Modified Dissolution Testing Apparatus 2

Shrutiben R. Parekh and **Piero M. Armenante**, Otto H. York Department of Chemical, Biological and Pharmaceutical Engineering, New Jersey Institute of Technology, Newark, NJ

Dissolution tests are routinely carried out in the pharmaceutical industry to determine the dissolution rate of solid dosage forms. Dissolution testing serves as a surrogate for drug bioavailability through in vitro–in vivo correlation (IVIVR), and it additionally helps in guiding the development of new formulations and in assessing lot-to-lot consistency, thus ensuring product quality. The United States Pharmacopoeia (USP) Dissolution Testing Apparatus 2 is the device most commonly used for this purpose. Despite its widespread use, dissolution testing using this apparatus remains susceptible to significant error and test failures. There is documented evidence that this apparatus is sensitive to several geometric variables that can affect the release profile of oral dosage forms, including tablet location during the dissolution process.

In this work, the dissolution profiles of disintegrating calibrator tablets containing Prednisone were experimentally determined using two systems, i.e., a Standard USP Dissolution Testing Apparatus 2 (Standard System) and a Modified Standard USP Dissolution Testing Apparatus 2 (Modified System) in which the impeller was located 8 mm off the vessel centerline. The dissolving tablets were located at different off-center positions on the vessel bottom to test the effect of tablet location in these two systems.

Tablet dissolution in the Standard System was found to be strongly dependent on tablet location, as previously reported by this and other research groups. This apparatus appears to generate variable results that may not be associated with the tablets undergoing testing but with the hydrodynamic characteristics of the apparatus itself and the location of the tablet on the vessel bottom. However, when the same experiments were conducted in the Modified System, the dissolution profiles for the same tablets were found to be nearly completely insensitive to tablet location.

The dissolution process in the Modified System was faster than that in the Standard System because of the improved mixing performance of the Modified System resulting from the non-symmetrical placement of the impeller. However, when the Modified System was operated at 35 rpm, the dissolution profiles for centrally located tablets were found to be very similar to those for the Standard System operating at 50 rpm. Unlike the Standard System however, the dissolution profiles obtained at 35 rpm in the Modified System were found to be insensitive to tablet location.

It can be concluded that the newly proposed Modified System for dissolution testing is a simple and yet robust and valid alternative to the current dissolution testing practice using the Standard USP Dissolution Testing Apparatus.

See more of this Session: Application of Quality by Design In Drug Product Process Development II
See more of this Group/Topical: Topical I: Comprehensive Quality by Design in Pharmaceutical Development and Manufacture

(138e) QbD More Than Just An Acronym

Sean E. Mackey, Third Party Technology & Device Development, Abbott Laboratories, Abbott Park, IL

To borrow from Aristotle, the whole of quality by design (QbD) is more than the sum of its parts. It is more than target product profiles and critical quality attributes. More than critical process parameters and design spaces. More than process analytical technologies and control strategies. This is not to suggest that these individual elements are devoid of value. Rather, it is in this systematic approach that pharmaceutical companies can fully realize the benefits proffered by QbD. This presentation will provide a case study where a holistic QbD approach was applied to development of a blended powder-in-capsule oral dosage form and how this approach led to more efficient utilization of resources, reduction of costs associated with development and commercial scale-up, and ultimately better product and process understanding.

See more of this Session: Application of Quality by Design In Drug Product Process Development II
See more of this Group/Topical: Topical I: Comprehensive Quality by Design in Pharmaceutical Development and Manufacture

() Implementing Quality by Design in Pharmaceutical Salt Selection

Jeremy Merritt[1], Shekhar Viswanath[2] and Gregory Stephenson[1], (1)Eli Lilly and Co., Indianapolis, IN, (2)Chemical Product Research and Development, Eli Lilly and Company, Indianapolis, IN

Active pharmaceutical ingredients (API) typically fall into the category of weak electrolytes. That is they can be classified as molecules with weakly acidic or basic functional groups. In their neutral state, API's are often poorly water-soluble, which can limit the bioavailability of the compound. Salt formation is often a useful handle to not only increasing the solubility, but also to alter the physical properties (melting point, polymorphic landscape, etc). In general, strong acids are needed to form salts of basic API due to the relatively low pKa of amine functional groups. If a salt undergoes reformation of the neutral components, a disproportionation reaction is said to take place. This reaction is important in the solid-state as it determines the shelf life of the product but also in solution as the API experiences different pH regions of the stomach and intestine enroute to absorption. Here we will present current progress on a first principles theoretical model to quantify the risk of disproportionation, including the effects of excipients. Experimental results validating the model will also be presented. The model allows one to estimate the balance between solubility enhancement and risk of disproportionation, enabling different salts to be evaluated using quality by design principles.

See more of this Session: Application of Quality by Design In Drug Product Process Development II
See more of this Group/Topical: Topical I: Comprehensive Quality by Design in Pharmaceutical Development and Manufacture

(160a) QbD Principles for Packaging Design and Stability: A Regulatory Perspective

Sharmista Chatterjee, Ondqa/Cder/FDA, FDA, Silver Spring, MD

Quality by Design (QbD) is a science and risk based approach for pharmaceutical manufacturing and development. Enhanced product understanding gained from following a QbD based approach, is increasingly being used to select packaging design parameters and to determine optimum storage conditions.The aim of this presentation is to illustrate, using relevant examples, various QbD based approaches that may be implemented for packaging design, determination of storage conditions and shelf life estimation and to offer a regulatory perspective regarding considerations for successful implementation of such approaches throughout the product life cycle.

See more of this Session: Implementation of the QbD-Paradigm for-Packaging Design and Estimation of Product Shelf Life
See more of this Group/Topical: Topical I: Comprehensive Quality by Design in Pharmaceutical Development and Manufacture

(160b) Stability In a QbD Framework

Theodora Kourti, GSK, Hamilton, ON, Canada

Multivariate Data Based Empirical models can be used to predict future product performance in terms of Stability, and anticipate product shelf life. So far Quality by Design is being used to assess the quality of a product at release based upon manufacturing data but this concept can be extended to investigate stability patterns based on multivariate interactions of raw material characteristics, manufacturing process parameters, and packaging characteristics. The DoEs for Design Space determination could be planned to include stability studies. Process Understanding derived from such studies would lead to appropriate Control Strategy that takes into account stability requirements

See more of this Session: Implementation of the QbD Paradigm for Packaging Design and Estimation of Product Shelf Life
See more of this Group/Topical: Topical I: Comprehensive Quality by Design in Pharmaceutical Development and Manufacture

(160c) Science-Based Stability and Packaging Development Under the Umbrella of QbD

Rey Chern, Pharmaceutical Commercialization Technology, Merck & Co, Inc., West Point, PA

Abstract

Science-Based Stability and Packaging DevelopmentUnder The Umbrella of QbD

Rey T Chern

Global Pharmaceutical Commercialization

Merck Sharp & Dohme

Asystematic approach for characterization of product stability risks, selectionof primary packages and utilization of first-principle modeling to managestability outcomes is presented. Theoverarching framework of this approach focuses on three areas. First, controlled Product CharacterizationStudies (PCS) are performed to gain quantitative understanding of the impact ofthe factors including environmental conditions (e.g. temperature, relativehumidity, oxygen, and light), formulation ingredients (e.g. API form, critical excipient impurities, etc.), and processing parameters (e.g.mechanical stress, drying, etc.) on the stability of the product.

Second,the primary containers under consideration are characterized quantitatively fortheir protective performance, specifically relevant to the stability drivers(e.g. moisture or oxygen transmission rate). Often, the critical environmental factorscontrolling the stability of oral products are temperature & water moisture. A model for quantitatively describing thetime-course of water activity that the product is exposed to over the expectedshelf life at the ICH testing conditions has been developed. This model, labeled as "**S**imulationof **H**eadspace **M**oisture **A**ctivity in **P**ackages(SHMAP)", is derived from the first principles of thermodynamics, mass balance,and mass transport dynamics. It has beenverified by experimental data collected for >20 development projects in thepast 10 years.

Third,the knowledge gained from PCS and SHMAP is combined to predict stability performanceof the various product/primary-container combinations intended for registration. This approach allows for improved design ofregistration stability studies and enables projection of thestability-performance of new product-packaging configurations with a highdegree of assurance since all needed scientific data input are available andcontrolled independently.

Thisapproach for stability & primary package development is a good example ofthe application of

Quality by Design principles.

See more of this Session: Implementation of the QbD Paradigm for Packaging Design and Estimation of Product Shelf Life
See more of this Group/Topical: Topical I: Comprehensive Quality by Design in Pharmaceutical Development and Manufacture

(160d) Using QbD to Asses the Factors That Affect Stability and Packaging

Stephanie Krogmeier, Regulatory Affairs, Vertex Pharmaceuticals, Cambridge, MA and Kevin J. Bittorf, CMC Process & Alliance Management, Strategic Development Networks, Vertex Pharmaceuticals, Cambridge, MA

Packaging is often the last step in drug product (DP) development; however packaging is extremely important in determining the marketability of a product. The ability to predict whether a product can be stored at room temperature early in development can affect how the drug is developed. In order to determine packaging requirements, and estimate shelf-life of the drug product it is important to understand the factors that impact DP degradation, which include storage temperature, water content, and starting materials.

Using quality by design (QbD) principles, an experimental plan and subsequent model were developed to systematically evaluate these factors. From the developed stability model, a strategy was determined to facilitate packaging selections, storage conditions and justification of specifications. The ability to use this model to determine the product packaging and justify the product specifications is crucial to patient safety and critical for a successful regulatory filing. Finally, discussed are the regulatory implications and how this information can be used in conjunction with ICH stability studies.

See more of this Session: Implementation of the QbD-Paradigm for-Packaging Design and Estimation of Product Shelf Life
See more of this Group/Topical: Topical I: Comprehensive Quality by Design in Pharmaceutical Development and Manufacture

(160e) Designing Pharmaceutical Packaging Using a Product-Packaging Stability Modeling

Seung-yil Yoon, Eli Lilly and Company, Indianapolis, IN

DesigningPharmaceutical Packaging using a Product-Packaging Stability Modeling

ABSTRACT

Pharmaceutical solid oral dosage forms are degradedby environment hazards such as moisture, oxygen, and temperature as a functionof time. One of functions of packagingis to protect products from those environment hazards. When a package is designed, variables such asproduct degradation kinetics, package barrier, sorbentsorption and environment conditions should be understood and integrated. These variables could be understood using a traditionaltrial and error approach where products are put in various combinations ofpackaging on stability screening for 6-12 months. However, it is time-consuming, costly and it can not be usedto explain a variety of unique situations that can arise during development andmanufacturing. Product-packagingmodeling integrates experimentally-determined product and packaging data, and themodeling provides estimated product stability throughout dynamically changingconditions inside the package during shelf life storage.

In this presentation, theories and key data inputsfor the product-packaging modeling will be explained and several successfulmodeling applications in support of QbD paradigm willbe shared.

See more of this Session: Implementation of the QbD-Paradigm for-Packaging Design and Estimation of Product Shelf Life
See more of this Group/Topical: Topical I: Comprehensive Quality by Design in Pharmaceutical Development and Manufacture

(160f) Implementation of QbD Paradigm In Sterile Dosage Form Packaging - Some Practical Considerations

Dr. Hemant N. Joshi, Tara Innovations LLC, Parsippany, NJ

Out of all the pharmaceutical dosage forms, the Quality Target Product Profile (QTPP) for the sterile dosage form is the strictest for obvious reasons. One has to pay attention to the critical quality attributes (CQA) to a great extent. This presentation is focused only on the packaging of sterile dosage forms. There are several types of sterile dosage forms such as small and large volume parenteral liquids, lyophilized products, ophthalmic solutions/suspensions etc. The packaging is different for each type. The Quality by Design (Qbd) approach should be applied during the packaging operation (identification, management and control of packaging operation). Leachables and extractables are critical in other dosage forms, but are most important in sterile dosage forms. QbD should also consider other aspects including – uniformity of distribution, pH, sterility, particulate matter, water content, antimicrobial and antioxidant preservative content, osmolarity, and endotoxins/pyrogens. In this seminars topics such as development of design space for packaging, influence of scaling up, design of clean room, environmental monitoring and aseptic processing will also be discussed. Important, design and interpretation of media fills will also be discussed to build in quality by design.

See more of this Session: Implementation of the QbD-Paradigm for-Packaging Design and Estimation of Product Shelf Life
See more of this Group/Topical: Topical I: Comprehensive Quality by Design in Pharmaceutical Development and Manufacture

(202a) Application of Design Space Mapping Accounting for Model Uncertainty and Common Cause Variability In Drug Substance Development

Shankar Vaidyaraman, Isabel Figueroa, Sze Wing Wong and Shekhar Viswanath, Chemical Product Research and Development, Eli Lilly and Company, Indianapolis, IN

This talk will focus on the methodology for mapping probabilistic design space based on first principles models and illustrate it with applications within drug substance development. The uncertainties in a model can be classified into two categories (1) Uncertainty in model parameters fitted from experimental data (2) Common cause variability in input factors of the model. The incorporation of these uncertainties present some statistical and computational challenges in the context of first principles models which will be highlighted in the talk. The application of design space mapping will be illustrated using case studies on reaction kinetics and reactive distillation unit operations.

See more of this Session: Application of Quality by Design In API Process Development
See more of this Group/Topical: Topical I: Comprehensive Quality by Design in Pharmaceutical Development and Manufacture

(202b) Design Space Development for the Final Recrystallization Process for Axitinib, a Complex Solvating API

Steven M. Guinness, Engineering Technologies - Chemical Research & Development, Pfizer Inc., Groton, CT, Brian P. Chekal, Chemical Research & Development, Pfizer Global Research & Development, Groton, CT and Gregory S. Steeno, Non-Clinical Statistics, Pfizer Global Research & Development, Groton, CT

This example for axitinib, an oncology candidate recently submitted to the FDA, will examine the development of the final processing step (a seeded recrystallization) to deliver final drug substance to meet the desired solid form and quality specifications. The process development work for the recrystallization is a textbook case of the use of sequentially designed multivariate experiments coupled with first principles to establish a robust design space.

See more of this Session: Application of Quality by Design In API Process Development
See more of this Group/Topical: Topical I: Comprehensive Quality by Design in Pharmaceutical Development and Manufacture

(202c) Impurity Fate Mapping In the Development of Pazopanib

Michael A. McGuire, Synthetic Chemistry, Glaxo SmithKline, King of Prussia, PA

Pazopanib hydrochloride is a VEGF antagonist developed by GSK which was approved for the treatment of renal cancer by the FDA and EMEA in 2009 -2010. Quality by design principles were used in the development of pazopanib hydrochloride. Impurity fate mapping was used extensively to understand and control impurities. This paper will discuss the impurity fate mapping strategy for pazopanib hydrochloride and how this strategy helps guarantee patient safety.

See more of this Session: Application of Quality by Design In API Process Development
See more of this Group/Topical: Topical I: Comprehensive Quality by Design in Pharmaceutical Development and Manufacture

(202d) A Model-Centric Solution to Link Content Uniformity Targets with API Particle Size Specifications and Process for a QbD Excercise

Salvador Garca-Muoz[1], Weili Yu[2], Mark A. Pinto[3] and Sean K. Bermingham[3], (1)Process Modeling & Engineering Technology, Pfizer Worldwide Research & Development, Groton, CT, (2) Pharmaceutical Development, Pfizer Worldwide Research & Development, Groton, CT, (3)Process Systems Enterprise, London W6 7HA, United Kingdom

One of the contributing factors in determining targets for the particle size of an API can be the content uniformity of the dosage forms. This work presents a framework established with a model-based specification for particle size of the API that can be used in a bi-directional mode to: a) determine center point processing conditions for the API given a target in content-uniformity for the dosage form, and b) to propagate expected sources of variability in the API process onto the final quality of the product.

This work discusses the necessary considerations and advantages when a model-based specification is considered for the particle size distribution of the API and how this compares to common practices involving discrete points of the size distribution (such as D[10], D[50], D[90]).

Our discussion is centered on the need of a structural specification for the API particle size that can be effectively used to guide the design of the last stages of the drug substance manufacturing train. We also emphasize the need to standardize the descriptors of particle size in order to link the effects of the variability in the API process as it affects the drug product train and eventually the quality of the dosage form.

See more of this Session: Application of Quality by Design In API Process Development
See more of this Group/Topical: Topical I: Comprehensive Quality by Design in Pharmaceutical Development and Manufacture

Process QbD Case Study: An Integrated PAT Approach for a Dynamic Pharmaceutical Co-precipitation Process Characterization and Process Design Space Development

Huiquan Wu*, Maury White, Mansoor Khan, Food and Drug Administration, Silver Spring, MD
(*Corresponding author: Huiquan.wu@fda.hhs.gov; Tel: 301-796-0022)

Abstract: Previously we had explored the concepts of developing an integrated PAT approach to monitor and map the state of a dynamic pharmaceutical co-precipitation process in real time, and to determine the nucleation and growth mechanisms occurred during a pharmaceutical co-precipitation process. The aim of current work was to develop a process QbD case study that demonstrates the feasibility of using an integrated PAT approach for a dynamic pharmaceutical co-precipitation process characterization and design space development at laboratory scale. A dynamic co-precipitation process by gradually introducing water to the ternary system of naproxen-Eudragit L100-alcohol was monitored at real-time *in situ* via Lasentec FBRM and PVM. The 3-D map of count-time-chord length revealed three distinguishable process stages: incubation, transition, and steady-state. The effects of high risk process variables (slurry temperature, stirring rate, and water addition rate) on both derived co-precipitation process rates and final chord-length-distribution were evaluated systematically using a 3^3 full factorial design. Critical process variables were identified via ANOVA for both transition and steady state. General Linear Models (GLM) were then used for parameter estimation for each critical variable. Clear trends about effects of each critical variable during transition and steady state were found by GLM and were interpreted using fundamental process principles and Nyvlt's transfer model. Neural network models were able to link process variables with response variables at transition and steady state with R^2 of 0.88~0.98. PVM images evidenced nucleation and crystal growth. Therefore, this work illustrated the possibility of combining first-principle modeling, mechanistic modeling, and empirical modeling approaches for dynamic process characterization and process design space development. Secondly, the technical feasibility of using an integrated PAT approach for in-depth process understanding and process design space development was demonstrated. Finally, the regulatory science perspective around the core concepts was briefly discussed.

1. Introduction

According to the definition of ICH Guideline Q8(R2) (FDA/ICH, 2009), "Quality-by-Design (QbD) is a systematic approach to development that begins with predefined objectives and emphasizes product and process understanding and process control, based on sound science and quality risk management." During the past few years, the pharmaceutical quality (Woodcock, 2004) and QbD implementation strategies (Yu, 2008; Rathore and Winkle, 2009; Smith et al., 2009) for various pharmaceutical product areas have received a lot of attentions. The FDA's PAT Guidance (FDA, 2004) and the ICH Q8, Q8(R2), Q9, and Q10 guidelines (FDA/ICH, 2006a,b, 2007, 2009) have collectively set up a roadmap for the implementation of the science-based and risk-based regulatory pathway. This can be characterized as pharmaceutical quality paradigm shift: from testing into quality alone to building the quality into the process and product.

In the QbD framework, understanding the impact of raw material attributes and process parameters on the critical quality attributes (CQAs) as well as identification and control of sources of variability are essential. However, pharmaceutical products and processes are complicated and multivariate in nature. Understanding of the relevant multi-factorial relationships among formulation parameters, process variables, and product quality attributes usually requires the use of multivariate approaches, such as statistical design of experiments (DOE) and multivariate data analysis (MVDA). Although both DOE and MVDA as standard techniques have been widely used for many years in various areas, the integrated use of both of them are relatively few when applied to pharmaceutical development (Naelapaa et al., 2008; Huang et al., 2009).

Over the past few years, some internal efforts (Wu and Hussain 2005; Wu and Khan 2009; Wu et al., 2009; Xie et al., 2008) have been made in the aspect of integration of DOE and MVDA for pharmaceutical product and process understanding, process monitoring and control. In principle, this integrated approach in conjunction with appropriate real time PAT process analyzer should provide ample opportunities for enhanced process understanding, process monitoring, and process control of other pharmaceutical processes (such as crystallization, co-precipitation, etc.) under the QbD/PAT framework.

Previously (Wu & Khan, 2010, 2011) we had explored the concepts of developing an integrated PAT approach to monitor and map the state of a dynamic pharmaceutical co-precipitation process in real time, and to determine the nucleation and growth mechanisms occurred during a pharmaceutical co-precipitation process. A uni-variate thermodynamic approach had been used for those works. Three thermodynamically important formulation and process variables were studied, including the drug/polymer ratio, water usage, and alcohol usage, etc. It demonstrated that integration real-time PAT process monitoring with first-principles modeling and off-line product characterization could enhance understanding to co-precipitation process dynamics and nucleation/growth mechanisms, which is impossible via off-line techniques alone. The knowledge gained through the above studies is essential for rationale design of both the co-precipitation process and co-precipitation vehicle. Furthermore, they can provide scientific evidence as for setting up practical operation range during the stage of process and product development.

The aim of current work was to develop a process QbD case study that demonstrates the feasibility of using an integrated PAT approach for a dynamic pharmaceutical co-precipitation process characterization and design space development at laboratory scale. In this work, a variety of independent process conditions were studied via a 3^3 full factorial design. For each set of process conditions, the co-precipitation process was monitored in real time using an integrated PAT approach. The effects of high risk process variables (slurry temperature, stirring rate, and water addition rate) on both derived co-precipitation process rates and final chord-length-distribution were evaluated systematically. Various models such as ANOVA Models, General Linear Models (GLM), and Neural Network (NN) models were able to link process variables with response variables at transition and steady state. A co-precipitation process design space was established.

2. Experimental and Data Analysis

2.1. Materials and Experimental Protocol for a Dynamic Co-precipitation Process

Naproxen USP (lot No: NPX 368) was obtained from Albemarle Corporation (Orangeburg, SC). Eudragit L 100 (lot No: 1221203048) was obtained from Röhm America Inc. (Somerset, NJ). Solvent reagent alcohol (HPLC grade, lot No: 053546) was purchased from Fisher Scientific (Fair Lawn, NJ). Non solvent DI water was obtained from a FDA in-house facility (Silver Spring, MD) and was kept in refrigerator at 3^0C prior to use. All of these chemicals, solvent, and non solvent were used without any further processing or purification prior to the use for this experimental work.

In this work, the drug/polymer ratio of 2 (4.0g of naproxen and 2.0g of Eudragit L100) was selected due to the fact that nice co-precipitates with regular morphology could be obtained via the dynamic co-precipitation process designed in this work. The naproxen and Eudragit were dissolved separately in 150 and 100 mL of reagent alcohol, respectively. These solutions were then added to the 1L reaction vessel of Chemglass reaction kit (ChemGlass,Vineland, NJ) via a funnel. The stir speed was set via the OptiChem digital overhead stirrer. The temperature was set via the ThermoScientific Neslab RTE 740 thermoregulator. The solution in the reaction vessel was allowed to reach the pre-defined temperature prior to data acquisition. Once this temperature had been achieved, real-time monitoring began using the Lasentec FBRM system and PVM system, prior to any water addition. The first addition of water (125 ml) was made to the reaction vessel at the pre-defined flow rate via a flexible tubing (Tygon R-3603, Fisher Scientific), a MasterFlex solid state speed control device (Chicago, IL), and a Cole-Parmer peristaltic pump (Chicago, IL). Ten minutes later, the 2nd addition of water (50 ml) was made to the system. Depending on the temperature of the individual experiment (15^0C, 25^0C, or 35^0C), the system was introduced with water for the 3rd, 4th, and 5th addition (each time with 50 ml) respectively. The time interval between each addition of water was kept as 10 minutes. To ensure both the experimental consistency and good dispersion of water to the ternary system, the flexible tubing was kept vertically by two supporting claps attached to a lab frame. Furthermore, the end portion of the tubing was immersed in the liquid phase during the entire course of experiment.

2.2. Risk-Based Approach to Select Variables for the DOE Study

A risk assessment tool was applied to identify potential variables that can have an impact on the desired quality attribute. In this work, a Fishbone diagram or Ishikawa diagram was created to structure the process of identifying possible causes that could have impacts on co-precipitation process rate, as shown in Figure 1.

Although six categories of factors (materials, process, measurement, equipment, people, and environment) could have impact on the co-precipitation process rate, given a research laboratory setting, a large number of factors can actually be eliminated safely. For example, co-precipitation formulation components selection and pre-dissolution in solvent, co-precipitation vessel selection, errors due to operators and analytical, environmental variables, and measurement variables, could be pre-specified in the study protocol. Therefore, this risk mitigation and risk management strategy would allow us to focus on evaluating the impact of main process variables or high risk variables (such as slurry temperature, slurry stirring rate, and non solvent addition rate) on process dynamics. Furthermore, this simplified situation will enable us to exemplify the concept proposed in this work.

2.3. Design of Experiments (DOE), Real-time Process Monitoring, and Data Analysis

Based on risk assessment results above, three main process variables (slurry temperature (^{0}C), slurry stirring rate (RPM), non solvent addition rate (ml/sec) were selected as independent process variables. Each factor was tested with three levels. A 3^3 full factorial design was used for experiments as shown in Table 1.

A FBRM D600L (Mettler-Toledo AutoChem, Columbia, MD) system was used to monitor the dynamic co-precipitation process *in situ* at real-time for a total of 27 batches based on a 3^3 factorial design. The FBRM D600 L system was set to collect Chord-Length-Distributions (CLDs) *in situ* with a laser speed of 2 m/s, applying the standard 'F-electronic' mode and using measurement duration of 2 seconds. A Lasentec Particle Vision Microscopy system PVM819 (Mettler-Toledo AutoChem, Columbia, MD) was used to monitor the co-precipitation process *in situ* at real-time for the selected batches. The PVM image was acquired every 2 seconds.

A 3-D count-time-size map was constructed. It was used to illustrate the progress of a co-precipitation process and to identify various distinguishable process stages. The impacts of main process variables on the process characteristics such as derived co-precipitation process rates and final CLD were assessed. ANOVA was conducted to identify statistical significant variables. General Linear Models were used for parameter estimate. Neural network algorithm was utilized to link the process variables with response variables.

3. Results and Discussion

From a process engineering perspective, various process variables could impact thermodynamics, kinetics, and mass-transfer of a co-precipitation process to various extents, depending on the combination of process conditions. Therefore, the effects of main process variables on the co-precipitation process dynamics and quality attributes of final co-precipitate slurry were examined systematically. The following chord length (CL) range codes were used throughout this paper: CL1: 0-10 μm; CL2: 10-19.953 μm; CL3: 19.953 -50.119 μm; CL4: 50.119-100 μm; CL5: 100-251.189 μm.

3.1. 3-D Count-Time-Size Plot of the Dynamical Co-precipitation Process

A 3-D map of count-time-size was constructed to visualize the co-precipitation process dynamics as a function of time. It can illustrate the process progress and distinguish various process stages (Wu and Khan 2010a), as shown in Figure 2 for process run 132.

It was found for most runs, the 3-D maps consist of three stages: (1) incubation period when there was no significant particle counts detected, its time length depending on the combinations of process conditions; (2) transition period when significant nucleation and growth occurs thus significant counts were detected; (3) steady state when the number of counts per second for small chord length ranges (CL1 and CL2) remains relatively constant throughout the remainder of the process, which indicates no significant nucleation occurs. In the CL3 range, there is a slight increase after the 2nd addition of water and a slight decrease after the 3rd addition of water for run 132; there is a slight decrease after the 2nd and 3rd addition of water for run 123.

3.2. Derived Process Rates for a Co-precipitation Process During the Transition Period

Although CLD is a relative value, it can be used to represent the process characteristics of a particulate system because it depends on both PSD and shape of the particles (both are a function of process time). Based on the profile of the FBRM counts/s vs. time at specific CL range during the transition period, a new procedure was developed in this work to derive the co-precipitation process rates from FBRM counts/s vs. time profiles. The slopes were based on the FBRM counts/s over time in five different CL ranges. They were calculated by (1) first identifying the general time range where the maximum slope occurred for a given size range during a single addition; (3) then calculating the slope of that CL range over the preceding 20 seconds for each individual time point within that CL range; (3) the maximum of these calculated slope values for each addition and each CL range was used as the derived co-precipitation process rate for the corresponding process conditions at specific single addition and CL range. Following this procedure, the slope of BBRM counts/s over time for each run after each addition of water was obtained. The following derived process rate codes were used throughout the paper: Rate1, Rate2, Rate3, Rate4 are corresponding to the derived rates at CL1, CL2, CL3, and CL4 after the 1st water addition; Rate5, Rate6, Rate7, Rate8 are corresponding to the derived rates at CL1, CL2, CL3, and CL4 after the 2nd water addition. These derived process rate data as the designated response variable values were listed in Table 2.

3.3. CLD at the End of Steady State Period

To characterize the effects of main process variables on the steady state, the CLDs at the end of steady state of the 27 runs were plotted. Figure 3 is a representative example for run 111, 112, and 113. All of the CLDs have bell shapes. As characteristic parameters of this bell- shaped CLD, its mode and peak frequency were taken as additional designated response variables and listed in the last section of Table 2.

3.4. Analysis of Variance (ANOVA) and General Linear Modeling for Both Transition Period and Steady State Period

Statistical data analysis and general linear modeling were performed on data from both transition period and steady state period.

3.4.1. ANOVA for the DOE Results of the Co-precipitation Process: Identifying critical process variables and interactions

Let X1, X2, and X3 be the slurry temperature, stirring rate, and water addition rate for the 3^3 full factorial design, respectively. With 27 runs of 3^3 factorial design of experiment, the analysis of variance (ANOVA) indicates that all main effects and two-way interactions among these main effects are estimable and the three-way interaction is used to estimate the underlying error. Therefore, the following statistical model was used:

$$y_{ijk} = \mu + \alpha_i + \beta_j + \gamma_k + (\alpha\beta)_{ij} + (\alpha\gamma)_{ik} + (\beta\gamma)_{jk} + \varepsilon_{ijk}$$

$$i = 1,2,3 \quad j = 1,2,3 \quad k = 1,2,3 \tag{2}$$

where y_{ijk} is the observation (either derived process rate value, mode, or peak frequency) in the ith level of variable X1 and the jth level of variable X2 and the kth level of variable X3, μ is the overall mean, α_i is the ith X1 effect, β_j is the jth X2 effect, γ_k is the kth X3 effect, $(\alpha\beta)_{ij}$ is the two-way interaction between the ith X1 effect and the jth X2 effect, $(\alpha\gamma)_{ik}$ is the two-way interaction between the ith X1 effect and the kth X3 effect, $(\beta\gamma)_{jk}$ is the two-way interaction between the jth X2 effect and the kth X3 effect, ε_{ijk}, the random error, is normally independently identically distributed with mean 0 and variance σ^2, which is written as $\varepsilon_{ijk} \sim iid\ N(0,\sigma^2)$.

This statistical model was used for ANOVA to screen significant factors and interactions. It reveals that, at significance level α=0.05, X1 and X3 are statistically significant for both Rate1 and Rate7; X1, X2, and X3 are statistically significant for Rate5, Rate6, Rate8, and peak frequency; X1 and X2 are statistically significant for mode. No statistically significant variables or interactions were found for Rate2, Rate3, and Rate4.

For the initial addition during transition period, only X1 and X3 stand out as critical process variables to the derived co-precipitation process rate for the smallest CL range of [0, 10] μm which is most likely related to the nucleation process. Since X1 is directly related to the degree of super-saturation (*S*), X3 is directly related to the creation of *S* in the co-precipitation vessel, the ANOVA results for the 1st water addition suggest that the initial nucleation process is thermodynamically controlled. No statistically significant process variables and interactions were found from ANOVA at other CL ranges after the 1st water addition. This suggests that the competition among various process mechanisms (such as thermodynamics for super-saturation creation, nucleation kinetics, and multi-phase mass transfer, etc.) is at such state that none of them becomes dominant.

For the 2nd addition during the transition period, all of three process variables (X1, X2, and X3) stand out as critical variables to Rate5, Rate6, and Rate8. This may suggest all of three process variables can contribute directly to the co-precipitation process rates at CL5, CL6, and CL8. X1 and X3 stand put as critical process variables to Rate7, which may suggest both the thermodynamic factor (X1) and a mass-transfer factor (X3) are playing equally important role.

At the end of steady state of the co-precipitation process, X1 and X2 are critical process variables to the mode of the final CLD. This suggests both the thermodynamic factor (X1) and mass-transfer factor (X2) are most influential to the particle size of the final co-precipitate slurry. X1, X2, and X3 are critical process variables to the peak frequency of the final CLD. This suggests that all of the main process variables directly contribute to the number of counts generated. Therefore, the ANOVA results for the characteristic parameters of the final CLD demonstrated the critical importance of process control (i.e., controlling process parameters as studied in this work) to achieve the desired product quality attributes at the end of the co-precipitation process.

3.4.2. Development of Statistical Predictive Models Using General Linear Modeling

Based on the screening results of statistically significant process variables, further statistical analyses were carried out using general linear models (GLM). The general linear models and parameters' estimations are listed in Table 3.

To illustrate the GLM model formulation, let us discuss the linear model for Rate1.

$$(Rate1)_i = \beta_0 + \beta_1 (X1)_i + \beta_3 (X3)_i + \varepsilon_i, i = 1,....,27, \quad (3)$$

where $(Rate1)_i$ is the ith observation of Rate1, $(X1)_i$ and $(X3)_i$ are the ith values for X1 and X3, respectively, error term ε_i is normally independently identically distributed with mean 0 and variance σ^2, which is written as $\varepsilon_i \sim iid\ N(0, \sigma^2), i = 1,...,27$. The GLM modeling was carried out using SAS9.2. The modeling results were summarized in Table 4. From the parameter estimate listed in the 5th and 6th columns, it is easily appreciated that the effects of main process variables on derived process rates, mode and peak frequency of the final CLD have certain trends as discussed below.

3.4.2.1. Slurry temperature (X1)

It has a significant effect (at significance level α=0.05) on the derived process Rate1, Rate5, Rate6, Rate7, and Rate8. Furthermore, since parameter β_1 estimate is negative for these rates, the lower slurry temperature will result in greater derived process rates. This could be explained from a thermodynamic point of view, based on the fact that both solubility and degree of super-saturation are function of slurry temperature.

On the other hand, X1 was a significant factor (at significance level α=0.05) on mode of final CLD at the end of steady state. However, parameter β_1 estimate is positive for mode. Therefore, the lower slurry temperature will result in smaller mode or finer co-precipitate crystal.

3.4.2.2. Slurry stirring rate (X2)

It was a significant factor (at significance level α=0.05) for Rate5 and Rate6. The parameter β_2 estimate is positive for Rate5, Rate6, and Rate8. This indicates that a higher X2 will result in a higher Rate5, Rate6, and Rate8. This trend is likely attributed to the possible mixing effect caused by stirring, as discussed below.

From a fluid dynamics perspective, local and mean values of the specific power input caused by the rotation speed of the impeller determine the degree of macro-mixing and micro-mixing. The bulk mixing inside the co-precipitation vessel must be ensured that its elements participate equally in the mass transfer process and hence the entire volume is sufficiently utilized. In addition to this bulk mixing, sufficient micro-mixing has to be present in the vicinity of contacting zone where water is introduced and fed to the slurry via the feed inlet, in order to even out quickly any local super-saturation peaks generated A lower X2 may not be strong enough to guarantee an effective mixing at both macro and micro level. On the other hand, a higher X2 could promote micro-mixing in the vicinity of contact zone thus reduce the super-saturation there. Furthermore, a higher X2 can facilitate mass transfer process including both bulk transfer and diffusion across the diffusion layer. Therefore, according to the Nyvlt's model (Nyvlt et al., 1985), the apparent overall derived process rates covering the three steps of multi-phase mass transfer process are faster.

X2 was a significant factor (at significance level α=0.05) for both mode and peak frequency of the final CLD at the end of steady state. However, the parameter β_2 estimate is negative for mode but positive for peak frequency. This indicates that higher stirring rate will lead to smaller particles but more counts in its final CLD.

3.4.2.3. Addition rate of water (X3)

It has a significant effect (at significance level α=0.05) on Rate1, Rate5, Rate6, and Rate7, with a higher X3 resulting in a greater derived process rate, as indicated by the GLM modeling results (the parameter β_3 estimate is positive). Similar trend was observed for the derived process rate 8 although it did not meet the significance criteria. This trend could be explained by the generation of super-saturation and the re-distribution of various species in the vicinity of contacting zone. Firstly, the addition of water into the slurry system will directly change the concentrations of various species, which in turn will change the degree of super-saturation in the vicinity of contacting zone; secondly, it indirectly impacts the re-distribution of various species within the vessel. The generation rate of super-saturation in the vicinity of contacting zone due to a faster X3 could be larger than the consumption rate of the super-saturation due to micro-mixing caused by the stirring. Therefore, a trend of a higher X3 leading to a greater derived process rate was observed.

Similar trend was observed for peak frequency of the final CLD at the end of steady state of the co-precipitation process, although it did not meet the significance criteria.

3.5. Neural Network Modeling for Both Transition Period and Steady State Period

As revealed by ANOVA screening results, there are certain two-way interactions existing, although not necessarily statistically significant at significance level α=0.05. Therefore, there is certain nonlinearity for the co-precipitation process. A neural network (NN) algorithm implemented in JMP 7.0 software was utilized to construct the linkage between the three normalized process variables and response variables that include derived process rates at transition stage and mode and peak frequency of CLD at the end of steady state. Our modeling results show that the NN was able to generate quantitative models with decent correlation coefficients, as shown in Table 4. Most of the R^2 values are pretty high (0.88~0.98), Furthermore, the predicted response variable values agree well with the actual response variable values, as shown in Figure 4 for Rate1.

In summary, our NN modeling results suggest that quantitative correlations between the process variables and the derived process rate based on Lasentec FBRM CLD vs. time profile do exist. Therefore, it is possible to utilize these quantitative correlations for better process design and process control.

3.6. PVM Images Evidenced Nucleation and Crystal Growth during the Dynamic Course of Adding Water

In this work, 6 co-precipitation runs were selected for in-situ real-time monitoring using PVM during the entire course of co-precipitation process. First of all, it demonstrated that for most cases, no appreciable sizes of particles were detected via PVM before and immediate after the 1st addition of water to the ternary system. However, about 10 minutes after the 1st addition of water, right before the 2nd addition of water, co-precipitates with appreciable sizes were detected via PVM. This is qualitatively consistent with what FBRM data revealed that relative fewer counts of fine particles were detected after the 1st addition of water. Secondly, significant increases were observed via PVM for both the number and dimensions of the co-precipitates before and right after the 2nd addition of water. Again, this is qualitatively consistent with what FBRM indicated that significant counts at various CL ranges were detected after the 2nd addition of water. Further analysis demonstrated the trends of the dimension changes of co-precipitates at both length and width dimensions during the course of a co-precipitation process.

3.7. Discussion from a Regulatory Science Perspective

ICH Q8(R2) (FDA/ICH, 2009) defines the design space as "The multidimensional combination and interaction of input variables (e.g., material attributes) and process parameters that have been demonstrated to provide assurance of quality. Working within the design space is not considered as a change. Movement out of the design space is considered to be a change and would normally initiate a regulatory post-approval change process." In addition to identifying the main process variable effects, multi-factorial DOE study as conducted in this work permits the quantification of variable interactions. The information on the presence or absence of interactions might be valuable for process optimization, identifying cause of variability, or informing a failure modes and effects analysis (FMEA) for defining a design space. As a proof-of-concept, this simplified case study was mainly focused on the effects of process variables while keeping other variables fixed based on risk analysis and initial formulation development results. It is important to keep in mind that other variables such as formulation variables can certainly play vital roles during process characterization and design space development. Based on the GLM modeling results, both 2-D and 3-D contour plots were made for certain response variables of the dynamic co-precipitation process, such as Rate 7 for the transition period as shown in Figure 5 (a)-(b). Figure 5(a)-(b) demonstrated that maximum Rate 7 was obtained around the following process window: $-1<X1<0.1$, $-0.25<X3<0.95$.

As demonstrated in this and previous works, process knowledge and in-depth process understanding are essential to establish appropriate process control strategy and operational space, such that co-precipitation product with desired quality attributes and properties (such as PSD, morphology, etc.) could be achieved eventually. Therefore, they are key elements for developing appropriate process design space. Risk analysis, identification of critical formulation/process variables and interaction, understanding the effects of these critical variables and interactions on the key product quality attributes are essential steps to achieve enhanced process understanding. Validating and challenging design space are important steps to ensure a robust design space developed for a pharmaceutical manufacturing process. In this work, various PAT tools and QbD principles had been integrated together to generate a process QbD case study at laboratory scale. Development of other approaches and effects of scale on this dynamic pharmaceutical co-precipitation system are ongoing.

4. Conclusions

In this work, a dynamic pharmaceutical co-precipitation process of gradually introducing water (non solvent) to the ternary system of naproxen (durg) - Eudragit L100 (polymer) - alcohol (solvent) was monitored in real-time and *in situ* by using Lasentec FBRM and PVM technology. Based on the Fishbone analysis result, the impact of high risk process variables such as co-precipitation slurry temperature, slurry stirring rate, and non solvent addition rate on both the FBRM counts/s vs. time profile and chord-length-distribution(CLD) at the end of co-precipitation process were examined systematically, using a 3^3 full factorial design. A 3-D count-time-size plot was constructed to map the entire co-precipitation process and distinguish various process stages: incubation, transition, and steady state. A new method was developed to extract a derived process rate from the Lasentec FBRM counts/s vs. time profile by calculating the maximum slope of the FBRM conuts/s vs. time after each addition of water. Critical process variables were identified via ANOVA for both transition and steady state periods. General Linear Models (GLM) were used for parameter estimate of each critical variable. The GLM results demonstrated clear trends about the effects of each critical variable, which can be interpreted using principles of fundamental process phenomena and Nyvlt's diffusion layer model. Furthermore, contour plots demonstrated the process windows bordered by ranges of critical process variables where maximum Rate 7 and maximum mode can be obtained at the transition period and at the end of steady state, respectively. Neural network models were able to link process variables with response variables both at transition period and steady state with R^2 of 0.90~0.98. PVM images evidenced nucleation and crystal growth during the dynamic co-precipitation process.

Acknowledgements

The following persons and organizations are acknowledged: Dr. Meiyu Shen in CDER of FDA for strong statistical supports on DOE analysis and statistical modeling; Dr. Vincent Vilker at the Office of Testing and Research, CDER, FDA for insightful internal review;Mr. Eric Dycus at the Mettler Toledo AutoChem Inc.(Columbia, MD) for technical support to our Lasentec systems; Dr. San Kiang at the Bristol-Myers Squibb Company (New Brunswick, NJ), Mr. Steve Ware and Mr. Brendan Simon at the Chemglass Inc. (Vineland, NJ) for their technical discussion and assistance to our crystallization system; Dr. Richard D. Braatz at MIT (Boston, MA) for technical discussion on topics of crystallization;The FDA BioSciences Library for literature supports.

Disclaimer: ***The views and opinions expressed in this paper are only those of the authors, and do not necessarily reflect the views or policies of the FDA***.

References

FDA, 2004. "Guidance for Industry. PAT—A Framework for Innovative Pharmaceutical Development, Manufacturing, and Quality Assurance". Available at: http://www.fda.gov/cder/guidance/6419fnl.pdf

FDA/ICH, 2006a. Guidance for Industry. Q8 Pharmaceutical Development. Available at: http://www.fda.gov/cder/guidance/6746fnl.pdf

FDA/ICH, 2006b. "Guidance for Industry. Q9 Quality Risk Management". Available at: http://www.fda.gov/cder/guidance/7153fnl.pdf

FDA/ICH, 2007. Q10. Pharmaceutical Quality System. Draft Consensus Guideline. Available at: http://www.fda.gov/cder/guidance/7891dft.pdf

FDA/ICH, 2009. Guidance for Industry, Q8(R2) Pharmaceutical Development. Draft available at: http://www.fda.gov/downloads/Drugs/GuidanceComplianceRegulatoryInformation/Guidances/ucm073507.pdf

Huang, J., Kaul, G., Cai, C., Chatlapalli, R., Hernandez-Abad, P., Ghosh, K., Nagi, A., 2009. Quality by design case study: An integrated multivariate approach to drug product and process development. Intl J Pharm. 382, 23-32.

Naelapaa, K., Alleso, M., Kristensen, H.G., Bro, R., Rantanen, J., Bertelsen, P., 2008. Increasing process understanding by analyzing complex interactions in experimental data. J. Pharm. Sci. 98, 1852–1861.

Nyvlt, J., Sohnel, O., Matuchova, M., Broul, M., 1985. The Kinetics of Industrial Crystallization. Elsevier, New York.

Rathore, A.S. and Winkle, H., 2009. Quality by design for biopharmaceuticals. Nature Biotechnology 27(1), 26-34.

Smith, J., Moore, C., Nasr, M., 2009. Quality Trio Takes Final Shape. Pharmaceutical Technology 33(6), 66.

Tavare, NS. 1987. Batch Crystallizations: A Review. Chem. Eng. Comm. 61:259-318.

Woodcock, J., 2004. The Concept of Pharmaceutical Quality. American Pharmaceutical Review. 7 (6), pg. 10-15.

Wu, H. and Hussain, A.S., 2005. Integration of multivariate statistics and design of experiments to identify critical process variables for pharmaceutical process analytical technology (PAT) applications. In: American Statistical Association 2005 Proceedings of Joint Statistical Meetings, Minneapolis, Minnesota, August 7-11, 2005. CD-ROM , Mira Digital Publishing.

Wu, H. and Khan, M.A., 2009. Quality-by-Design (QbD): An Integrated Approach for Evaluation of Powder Blending Process Kinetics and Determination of Powder Blending End-Point. J Pharm Sci. 98(8), 2784-2798.

Wu, H., Tawakkul, M., White, M., and Khan, M.A., 2009. Quality-by-Design (QbD): An Integrated Multivariate Approach for the Component Quantification in Powder Blends. Int J Pharm. 372(1-2), 39-48.

Wu, H. and Khan, M.A., 2010. Quality-by-Design (QbD): An Integrated Process Analytical Technology (PAT) Approach for Real-time Monitoring and Mapping the State of a Pharmaceutical Co-precipitation Process. Journal of Pharmaceutical Sciences 99(3): 1516-1534.
Wu, H. and Khan, M.A., 2011. Quality-by-Design (QbD): An Integrated Process Analytical Technology (PAT) Approach to Determine the Nucleation and Growth Mechanisms during a Dynamic Pharmaceutical Co-precipitation Process.
Journal of Pharmaceutical Sciences 100(5): 1969-1986.
Xie, L., Wu, H., Shen, M., Augsburger, L., Lyon, R.C., Khan, M.A., Hussain, A.S., Hoag, S.W., 2008. Quality-by-Design (QbD): Effects of Testing Parameters and Formulation Variables on the Segregation Tendency of Pharmaceutical Powder Measured by the ASTM D 6940-04 Segregation Tester. J. Pharm. Sci. 97(10),4485-97.
Yu, L., Lionberger, R.A., Raw, A.S., D'Costa, R., Wu, H., Hussain, A.S., 2004. Applications of process analytical technology to crystallization processes. Advanced Drug Delivery Reviews. **56**(3): 349-369.
Yu, L.X., 2008. Pharmaceutical Quality by Design: product and Process Development, Understanding, and Control. Pharmaceutical Research 25(4), 781-791.

Table 1 The 3^3 full factorial design for the co-precipitation study

Run No	Pattern	X1(^{0}C)	X2(RPM)	X3(ml/sec)
1	133	15	400	6.66
2	213	25	50	6.66
3	312	35	50	3.66
4	132	15	400	3.66
5	331	35	400	0.83
6	123	15	200	6.66
7	323	35	200	6.66
8	221	25	200	0.83
9	313	35	50	6.66
10	222	25	200	3.66
11	321	35	200	0.83
12	212	25	50	3.66
13	311	35	50	0.83
14	131	15	400	0.83
15	211	25	50	0.83
16	332	35	400	3.66
17	223	25	200	6.66
18	112	15	50	3.66
19	122	15	200	3.66
20	231	25	400	0.83
21	232	25	400	3.66
22	121	15	200	0.83
23	233	25	400	6.66
24	322	35	200	3.66
25	113	15	50	6.66
26	111	15	50	0.83
27	333	35	400	6.66

Table 2. Derived process rates during transition and final CLD parameters at the end of steady state for the 27 co-precipitation runs based on real-time FBRM data (CL1: 0-10 μm;CL2: 10-19.953 μm;CL3: 19.953-50.119 μm;CL4: 50.119-100 μm).

Run #	Pattern	During the transition period (Rate: counts/s/s)								At the end of steady state period	
		After the 1st addition of water				After the 2nd addition of water					
		Rate1	Rate2	Rate3	Rate4	Rate5	Rate6	Rate7	Rate8	Mode(μm)	Peak frequency(counts/s)
1	133	10.5736	10.7710	10.5172	1.6980	24.6889	22.4311	21.5892	3.7845	27.123	305.72
2	213	2.2180	3.9841	8.4474	0.8013	1.4502	3.1170	8.1841	2.4616	46.416	91.581
3	312	0.3826	0.4129	0.9374	0.6608	5.2500	2.9591	4.9887	2.3592	54.117	46.169
4	132	2.7907	3.4472	4.2661	1.1092	29.2702	28.5301	26.8955	3.8570	25.119	270.67
5	331	1.6878	1.2320	2.7800	1.3865	2.8786	1.8109	3.3692	1.5301	42.987	60.755
6	123	17.4587	14.5936	12.1450	0.9303	31.2612	38.0034	55.6426	5.0592	31.623	392.19
7	323	1.2708	2.1529	3.9911	2.0179	6.3940	7.0410	11.3126	4.9157	39.811	109.45
8	221	0.8667	1.2557	3.3192	1.2608	0.7273	0.9842	0.4499	1.5104	46.416	92.221
9	313	0.3736	0.6492	0.9462	0.7991	1.8012	2.9355	5.7506	2.0944	54.117	59.715
10	222	4.7627	5.3115	5.7942	0.3398	12.8510	17.5157	35.4949	9.5555	36.869	254.68
11	321	1.0028	1.4100	3.4853	1.6461	1.7586	3.0196	4.7413	2.2973	39.811	87.882
12	212	16.8696	28.8235	59.4750	4.7923	10.3219	10.9985	27.4774	4.0631	36.869	260.91
13	311	0.5023	0.5972	1.4905	0.9732	0.9189	1.2675	2.4570	1.2153	50.119	34.156
14	131	2.4071	3.3990	3.2181	0.5674	9.6399	9.4964	15.6298	3.2423	34.145	205.99
15	211	0.2351	0.3937	0.9423	0.2103	1.0915	1.8768	5.8460	2.7091	50.119	88.607
16	332	4.2665	5.7420	8.6329	2.5536	5.8941	6.4683	11.9191	4.6014	34.145	127.26
17	223	5.4219	6.9866	8.5837	0.4530	10.4084	15.8936	34.5608	7.6965	34.145	252.77
18	112	20.9567	30.1534	62.3363	8.2928	17.1882	18.1251	36.7147	5.7329	31.623	285.74
19	122	3.8570	0.1250	0.1682	0.0324	14.4302	22.9464	35.9995	9.6199	34.145	292.63
20	231	1.0080	0.6780	1.0058	0.7622	9.3001	10.1723	17.7417	5.2822	31.623	155.29
21	232	0.8753	0.5902	0.8450	1.0472	12.5681	14.7266	26.3863	7.0553	31.623	171.35
22	121	4.3244	6.5958	10.1978	1.0433	5.3310	6.1365	7.4282	2.6867	31.523	243.11
23	233	1.5458	1.4770	1.7359	0.4937	11.0354	14.8249	31.6042	7.8285	29.286	203.1
24	322	0.8057	1.4403	2.7305	1.3534	3.8065	4.0625	9.3274	3.4756	39.811	82.84
25	113	17.8998	26.7439	50.1879	5.4916	2.2388	2.7659	6.1132	2.1597	36.869	201.59
26	111	1.2927	2.2658	4.0293	0.5283	3.5820	4.1017	10.2035	3.0273	34.145	131.14
27	333	11.7237	16.2410	26.0520	3.6388	5.0155	4.0808	8.1521	4.4261	36.869	159.24

Table 3 Development of statistical predictive models using general linear modeling

Process Period	Response Variable	Model	Coefficient of Correlation, r	Parameter	Estimate	Standard error	t-Value	P-value
Transition period after the 1st water addition	Rate1	$(Rate1)_i = \beta_0 + \beta_1(X1)_i + \beta_3(X3)_i + \varepsilon_i, \varepsilon_i \sim N(0,\sigma^2)$	0.595	β_0	5.09	1.02	5.01	<0.0001
				β_1	-3.31	1.24	-2.66	0.0138
				β_3	3.06	1.24	2.46	0.0214
Transition period after the 2nd water addition	Rate5	$(Rate5)_i = \beta_0 + \beta_1(X1)_i + \beta_2(X2)_i + \beta_3(X3)_i + \varepsilon_i, \varepsilon_i \sim N(0,\sigma^2)$	0.751	β_0	8.93	1.14	7.85	<0.0001
				β_1	-5.77	1.39	-4.15	0.0004
				β_2	3.69	1.39	2.65	0.0143
				β_3	3.28	1.39	2.36	0.0273
	Rate6	$(Rate6)_i = \beta_0 + \beta_1(X1)_i + \beta_2(X2)_i + \beta_3(X3)_i + \varepsilon_i, \varepsilon_i \sim N(0,\sigma^2)$	0.748	β_0	10.23	1.29	7.96	<0.0001
				β_1	-6.61	1.57	-4.19	0.0003
				β_2	3.58	1.57	2.27	0.0328
				β_3	4.01	1.57	2.55	0.0180
	Rate7	$(Rate7)_i = \beta_0 + \beta_1(X1)_i + \beta_3(X3)_i + \varepsilon_i, \varepsilon_i \sim N(0,\sigma^2)$	0.631	β_0	17.26	2.19	7.87	<0.0001
				β_1	-8.57	2.68	-3.19	0.0039
				β_3	6.39	2.68	2.38	0.0256
	Rate8	$(Rate8)_i = \beta_0 + \beta_1(X1)_i + \beta_2(X2)_i + \beta_3(X3)_i + \varepsilon_i, \varepsilon_i \sim N(0,\sigma^2)$	0.511	β_0	4.23	0.42	10.15	<0.0001
				β_1	-0.68	0.51	-1.33	0.1956
				β_2	0.88	0.51	1.72	0.0994
				β_3	0.94	0.51	1.84	0.0786
At the end of steady state period	mode	$(\text{mode})_i = \beta_0 + \beta_1(X1)_i + \beta_2(X2)_i + \varepsilon_i, \varepsilon_i \sim N(0,\sigma^2)$	0.854	β_0	37.83	0.83	45.76	<0.0001
				β_1	5.86	1.01	5.79	<0.0001
				β_2	-5.64	1.01	-5.57	<0.0001
	Peak frequency	$(peak)_i = \beta_0 + \beta_1(X1)_i + \beta_2(X2)_i + \beta_3(X3)_i + \varepsilon_i, \varepsilon_i \sim N(0,\sigma^2)$	0.856	β_0	172.84	10.07	17.16	<0.0001
				β_1	-86.74	12.33	-7.03	<0.0001
				β_2	25.54	12.33	2.07	0.0498
				β_3	37.57	12.33	3.05	0.0057

Table 4 Neural Network models' performance summary

Process Period	Response variable	SSE	RMSE	SSE Scaled	RMSE Scaled	R^2
After the 1st water addition during the transition period	Rate1(counts/s/s)	19.7528	0.9267	0.4963	0.1469	0.9809
	Rate2(counts/s/s)	54.0231	1.5326	0.6703	0.1707	0.9742
	Rate3(counts/s/s)	151.4385	2.5660	0.4891	0.1458	0.9812
	Rate4(counts/s/s)	6.2611	0.5217	1.7930	0.2792	0.9310
After the 2nd water addition during the transition period	Rate5(counts/s/s)	86.5533	1.9399	1.2218	0.2305	0.9530
	Rate6(counts/s/s)	121.0679	2.2943	1.3498	0.2422	0.9481
	Rate7(counts/s/s)	587.4973	5.0540	2.9544	0.3584	0.8864
	Rate8(counts/s/s)	11.6128	0.7106	2.0667	0.2998	0.9205
At the end of steady state period	Mode(μm)	99.5207	2.0801	1.5845	0.2625	0.9391
	Peak frequency (counts/s)	14200.851	24.8481	1.5675	0.2611	0.9397

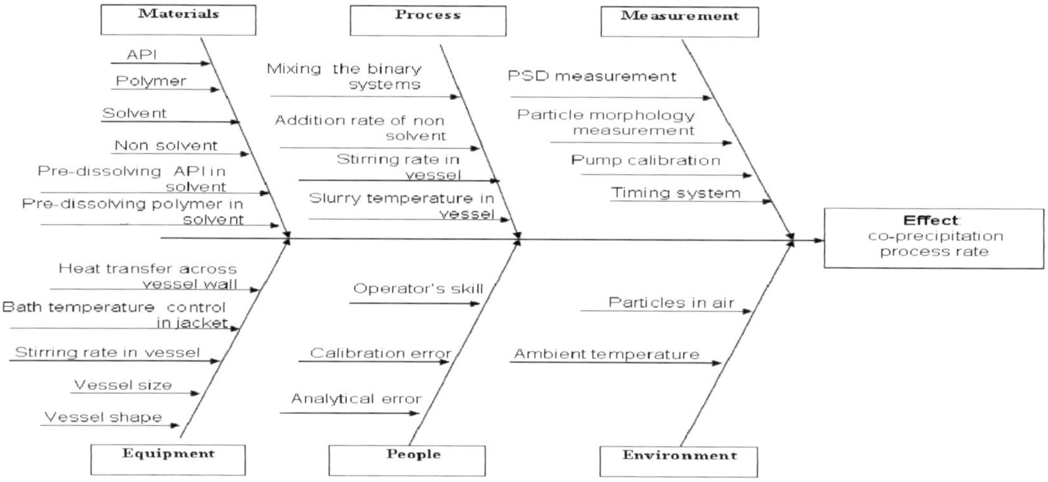

Figure 1.Fishbone diagram for risk analysis

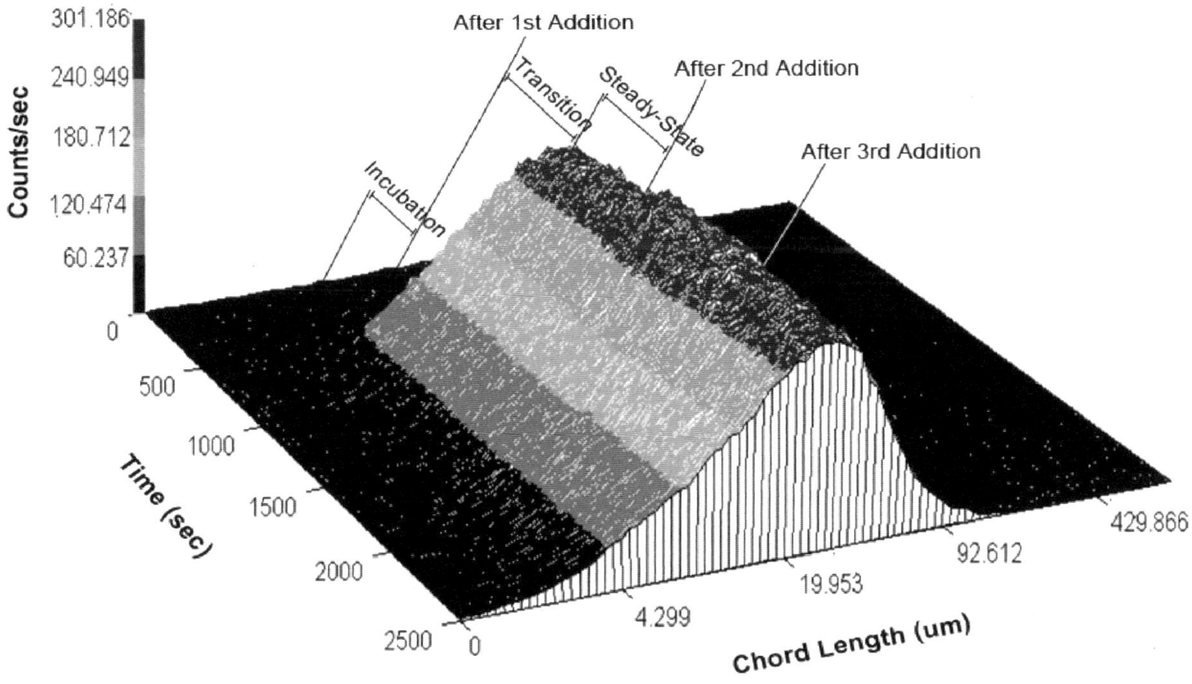

Figure 2. 3-D Count-Time-Size plots of a dynamic co-precipitation process run 132

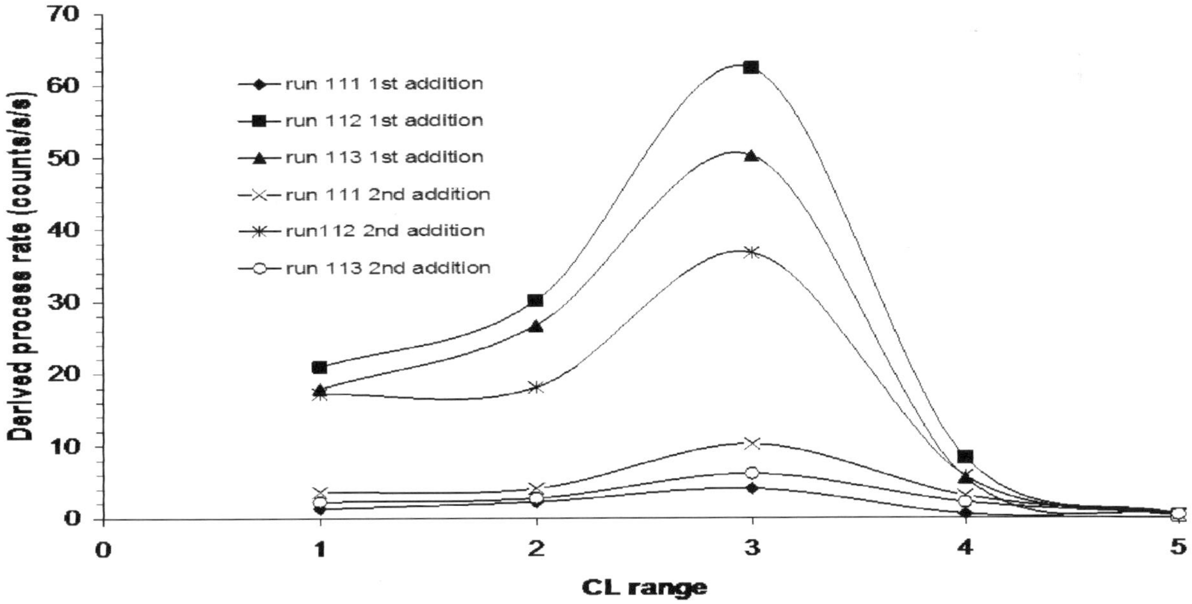

Figure 3. Plots of derived process rate vs. CL range

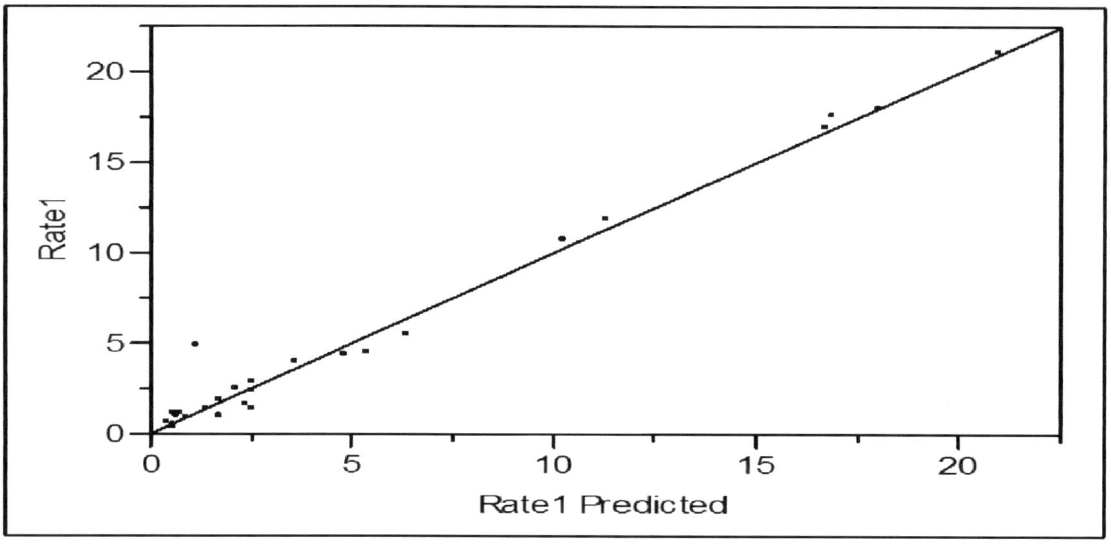

Figure 4. NN predicted response variable value vs. the actual response variable value for Rate1

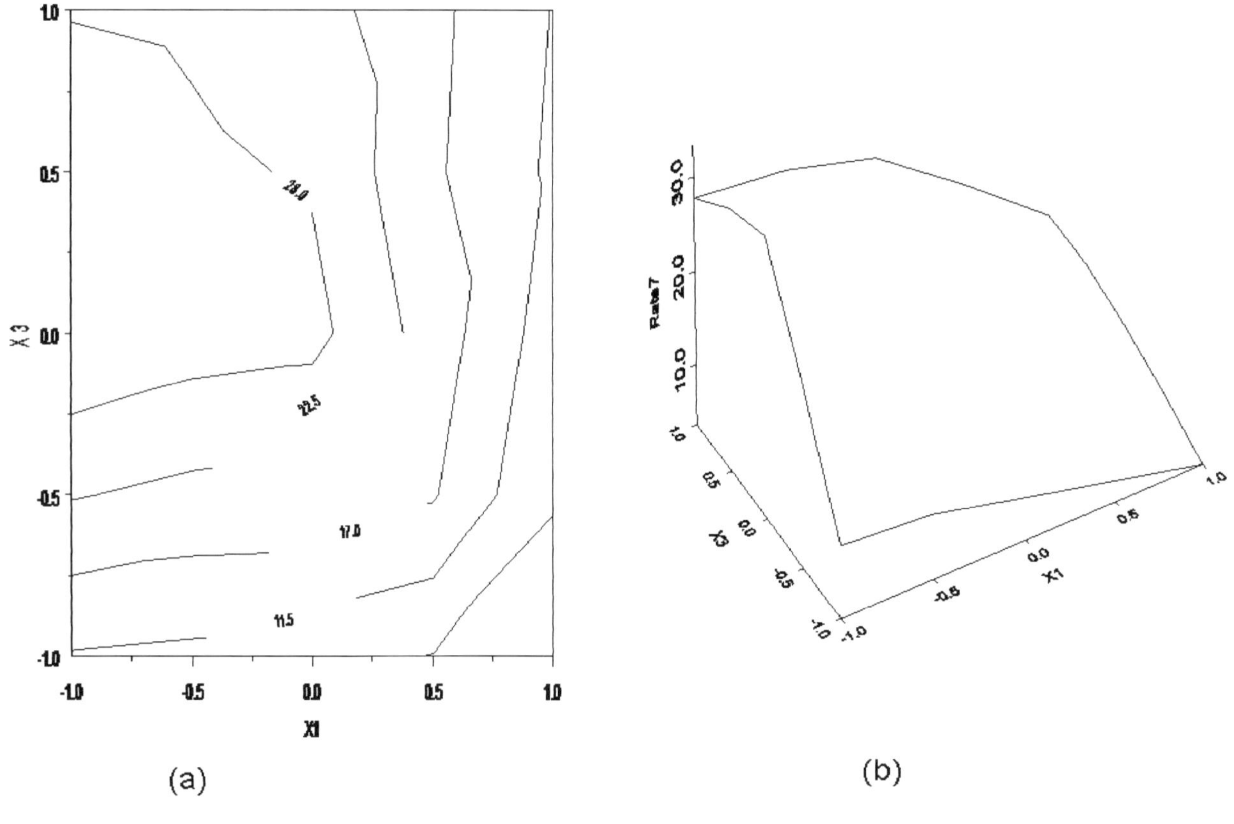

Figure 5. 2-D and 3-D contour plots for Rate7

(202f) A QbD Approach to Improve Tablet Coating Uniformity

Atul Dubey[1], Fani Boukouvala[1], Golshid Keyvan[1], Richard Hsia[2], Kostas Saranteas[2], Dean Brone[2], Tushar Misra[2], Marianthi Ierapetritou[1] and Fernando. J Muzzio[1], (1)Chemical and Biochemical Engineering, Rutgers University, Piscataway, NJ, (2)Chemical Process Research and Development, Sunovion Inc., Marlborough, MA

A combination of analytical and statistical methods is used to improve a tablet coating process guided by quality by design principles. A solid dosage form product was found to intermittently exhibit bad taste. A suspected cause was the variability in coating thickness which could lead to the subject tasting the active ingredient in some tablets. A number of samples were analyzed using a Laser Induced Breakdown Spectroscopy (LIBS) based analytical method and it was found that the main variability component was the tablet-to-tablet variability within a lot. Hence it was inferred that the coating process (performed in a perforated rotating pan) required optimization. A set of designed experiments along with Response Surface Modeling and Kriging method were used to arrive at an optimal set of operating conditions. Effects of the amount of coating imparted, spray rate, pan rotation speed and spray temperature were characterized. Results were quantified in terms of the relative standard deviation of tablet-averaged LIBS score and a coating variability index which was the ratio of the standard deviation of the tablet-averaged LIBS score and the weight gain of the tablets. The data driven models developed based on the designed experiments predicted that the minimum value of this index would be obtained for a 6% weight gain, for a pan operating at highest speed at the maximum fill level while using the lowest spraying rate and temperature from the chosen parametric space. This systematic application of QbD based method resulted in an enhanced process understanding and reducing the coating variability by more than half.

See more of this Session: Application of Quality by Design In API Process Development
See more of this Group/Topical: Topical I: Comprehensive Quality by Design in Pharmaceutical Development and Manufacture

(332a) FDA Perspective On QbD for Analytical Methods

Elaine Morefield, DHHS/FDA/Cder/OPS/ONDQA, FDA, Silver Spring, MD

QbD is a systematic approach to pharmaceutical development and product lifecyclemanagement. QbD provides new opportunities and modern tools to utilize in analytical research and development. ICH guidelines and implementation strategies focused mainly on formulation and manufacturing design and development. However, many of the QbD approaches and tools are applicable to analytical methods as well. The linkage between QbD and analytical method development and validation needs further development. Inclusion of enhanced pharmaceutical development information, including analytical method development and validation studies, in the submissions will provide the applicants the opportunity for flexible regulatory approaches, recognizing however, the degree of regulatory flexibility is predicated on the level of demonstrated product, analytical method and manufacturing process understanding. A summary of the progress made to date and the continued challenges facing the industry and FDA will be presented.

See more of this Session: QbD In Analytical Development
See more of this Group/Topical: Topical I: Comprehensive Quality by Design in Pharmaceutical Development and Manufacture

(332c) Quality by Design for Analytical Methods In Early Development

Jackson Pellett, Small Molecule Analytical Chemistry & Quality Control, Genentech Inc, South San Francisco, CA

In Phase 1 and 2 a high attrition rate and continuously changing targets can make spending resources of QbD activities seem very unattractive. However, applying many of the QbD principles early in the lifecycle can payoff with decreased method development time, fewer changes required due to poor method performance and a better understanding of method capability and risks. Column screening systems and modeling tools are often used in early development to identify and optimize separation conditions. Other tools such as risk assessments and method performance monitoring are less commonly used. Examples of early analytical development QbD approaches will be presented.

See more of this Session: QbD In Analytical Development
See more of this Group/Topical: Topical I: Comprehensive Quality by Design in Pharmaceutical Development and Manufacture

(332d) Examples of Applying Quality by Design Principles to Analytical Methods

Kimber Barnett, Analytical Research & Development, Pfizer Global Research & Development, Groton, CT, Timothy Graul, Analytical R&D US, Pfizer Global Research & Development, Groton, CT and Melissa Hanna-Brown, Dev Analy & ICH Stability, Pfizer Inc, Sandwich, Kent, United Kingdom

The application of Quality by Design principles to analytical method development practices is a science and risk based approach that begins with the identification of method requirements. Once method requirements have been identified, an analytical technique is selected based on these requirements. A risk assessment is then conducted to identify parameters that could potentially affect method performance. Subsequent experiments are conducted to develop a more complete understanding and control of sources of variability. Ultimately, this leads to the implementation of a robust measurement system that can support process evaluation and improvement throughout the product lifecycle. Examples that illustrate these concepts will be presented.

See more of this Session: QbD In Analytical Development
See more of this Group/Topical: Topical I: Comprehensive Quality by Design in Pharmaceutical Development and Manufacture

(332e) Quality by Design for Particle Size Analysis

Zhigang Sun, Office of Generic Drugs, OPS, CDER, FDA, Rockville, MD

Particle size analysis of pharmaceutical powders or dispersions is of great importance due to the potentially significant impact of particle size on safety, efficacy and quality of the drug product. The development of an appropriate particle sizing method is fundamental to establishing particle size specifications, which may be critical for control of drug product performance as well as ensuring manufacturing consistency. This presentation will describe a quality by design (QbD) approach to development of particle sizing methods. While QbD primarily focuses on product design and process development, it can be extended to development of analytical methods as well. In this approach, several QbD strategies (e.g., analytical target profile, method selection, risk assessment, control strategy, etc.) have been applied during the method development to ensure that the particle sizing method is rugged and robust and provides optimum performance. As a result, this more systematic approach could facilitate continual improvement throughout the method's lifecycle of use.

See more of this Session: QbD In Analytical Development
See more of this Group/Topical: Topical I: Comprehensive Quality by Design in Pharmaceutical Development and Manufacture

(407a) Progress In Quality by Design Implementation In Fda's Office of New Drug Quality Assessment

Christine Moore, DHHS/FDA/Cder/OPS/Ondqa, FDA, Silver Spring, MD

Since the launch of the Office of New Drug Quality Assessment (ONDQA) CMC pilot program in July 2005, over 40 Quality by Design (QbD) containing applications have been submitted, reviewed, and approved by ONDQA, both within and outside of the pilot. Although challenges still remain, many approaches for preparation and review of QbD containing applications have been established and efforts are underway toward international harmonization. This presentation will discuss the progress made to date in QbD and remaining scientific and regulatory challenges. Additionally, an overview will be given of harmonization efforts, including the recently announced pilot between FDA and the European Medicines Agency (EMA) for QbD containing applications.

See more of this Session: Round Table Invited Session On QbD: QbD Filings
See more of this Group/Topical: Topical I: Comprehensive Quality by Design in Pharmaceutical Development and Manufacture

(407b) QbD Approaches for Continuous Manufacturing

Fernando J. Muzzio, Chemical and Biochemical Engineering, Rutgers University, Piscataway, NJ

In this talk, we discuss multiple examples of application of QbD ideas and principles to the design of continuous manufacturing systems. Examples include the use of DoE methods for the characterization of system subcomponents (feeders, blenders, feed frames, tablet press) and the identification of critical process variables, examination of the effect of process parameters (blend composition, shear, strain) on the material properties of intermediate and finished products (blend uniformity and flowability, tablet hardness and dissolution) and the analysis of the integrated system.

See more of this Session: Round Table Invited Session On QbD: QbD Filings
See more of this Group/Topical: Topical I: Comprehensive Quality by Design in Pharmaceutical Development and Manufacture

(407c) Experiences with Quality by Design Filings and Prior Approval Inspections

Chris Balducci, Technical Research & Development, Novartis Pharmaceuticals Corporation, East Hanover, NJ

Experiences with Quality by Design Filings and Prior Approval Inspections

Abstract:

Quality by Design has been embraced by the Pharmaceutical Industry yielding much progress in interpreting ICH Guidelines, implementing risk assessment strategies and constructing Design Space philosophies in an effort to improve scientific understanding during product development. These activities are then assessed at the time of dossier filing, when years of development work are evaluated for alignment with Health Authorities expectations.

This short presentation will introduce the audience to Novartis' experiences regarding dossier module preparation and prior approval inspections conducted in the US and EU. Examples will be given highlighting the importance of source document quality and content and successful cross functional team interactions. The goal of this presentation is to provide enough detail to foster round table discussion and sharing of experiences regarding QbD filings.

See more of this Session: Round Table Invited Session On QbD: QbD Filings
See more of this Group/Topical: Topical I: Comprehensive Quality by Design in Pharmaceutical Development and Manufacture

(407d) Quality by Design for Generic Drugs

Yingxu Peng, Office of Generic Drugs, FDA CDER, Rockville, MD

To facilitate the implementation of Quality by Design (QbD) for abbreviated new drug application (ANDA), Office of Generic Drugs (OGD) has developed mock QbD examples for immediate release (IR) and modified release (MR) dosage forms. This presentation will provide insight into OGD's current thinking on QbD filing requirement and definition of some QbD nomenclatures. The Question-based Review (QbR) system is to be revised in 2011 and 2012 and is expected to be fully implemented by January 2013. This presentation will also discuss some practical tools and types of pharmaceutical development studies ANDA applicants may use as they implement QbD in their pharmaceutical development process.

See more of this Session: Round Table Invited Session On QbD: QbD Filings
See more of this Group/Topical: Topical I: Comprehensive Quality by Design in Pharmaceutical Development and Manufacture

QUALITY BY DESIGN (QBD), BIOPHARMACEUTICAL MANUFACTURE.

Kurt Brorson, Ph.D., Office of Biotechnology Products, CDER/FDA, 10903 New Hampshire Ave., Silver Spring MD 20903. Views expressed are those of the author and not necessarily those of the US FDA.

In the biopharmaceutical industry, consistently manufacturing a safe and high quality protein drug substance can be challenging. Recently, a regulatory structure supporting Quality by Design (QbD) for pharmaceuticals and biopharmaceuticals has been outlined in International Conference on Harmonization (ICH) Quality documents (Q8 through Q11; available at www.ich.org). Many of the concepts espoused in these ICH documents [e.g., quality target product profile (QTPP), risk-based approaches, product and process understanding, design spaces, control strategies with process analytical technologies (PAT), product life cycle, continuous process verification] are universal for pharma/biotechnology manufacturing and can readily be applied to bulk protein production.

The application of QbD to biotechnology development and manufacturing traces its origins to the early 2000's. In this timeframe, QbD concepts were being discussed and evolved during the implementation of CDER's GMP initiative and ICH discussions. These discussions identified the need for science-and risk-based approaches to improve quality and consistency of both traditional pharmaceuticals and biotechnology products. In 2005, CDER's Office of New Drug Quality Assessment (ONDQA) initiated a pilot program allowing sponsors to submit quality by design based applications (70 FR 40719, July 14, 2005). In 2008, CDER's Office of Biotechnology Products (OBP) follow suit, initiating a similar pilot program for biotech products (73 FR 37972, July 2, 2008). These applications were submitted to Federal Register Docket No. FDA-2008-N-0355, with a deadline to request participation by September 30, 2010. OBP's goal was to facilitate quality by design, risk-based approaches for manufacturing of complex molecules. As OBP consists of two divisions, the Division of Monoclonal Antibodies or the Division of Therapeutic Proteins, this program covers applications for monoclonal antibody, antibody fusion protein, cytokine, enzyme, hematologic proteins and other recombinant proteins (other than hormone or insulin products which are regulated by ONDQA).

To date, the program has successfully drawn applications from several firms. Applications accepted in QbD Pilot include five Original Applications (4 Monoclonal Antibodies and 1 Fc Fusion Protein) and four Post-approval Supplements (2 Monoclonal Antibodies, 1 Therapeutic Protein, and 1 multi-product). Two of these applications involve production site transfers; thus, OBP is working closely with CDER's Office of Compliance (OC). To enhance this cooperation, a CDER manual of policy and procedure (MAPP 4730.3) was recently developed to define and coordinate interactions between OBP & OC. The pilot program is intended to be highly interactive; on average, there are 3-5 meetings for each application/supplement. This level of interaction between CDER and regulated firms is critical due to the complexity of QbD applications and the need to proscriptively define goals, targets and agreements before initiating the large volume of work required in a QbD application.

Beyond the QbD applications submitted to CDER, hypothetical and real examples have been published. This is very useful from an educational standpoint because the submitted QbD dossiers are trade secret and thus the structure and detail is not publically available. In contrast, the hypothetical case studies are public, present a possible framework for QbD application and contain a high level of detail. The best known of these examples is "A-Mab: a case study in bioprocess development" written in 2009 by the CMC Biotech Working Group, a consortium of firms involved in biotech manufacture, with input from CDER (http://www.casss.org/associations/9165/files/A-Mab_Case_Study_Version_2-1.pdf). This case study is highly comprehensive, covering implementation of QbD from upstream manufacture to final fill/finish. It contains example risk analyses of individual unit operations and follow-up design of experiment studies. It also discusses aspects of QbD germane to molecule design, QTPP identification, linkage analysis, control strategy, life cycle management and regulatory. Two other case studies have been published which are restricted to one aspect of one unit operation, viral clearance by anion exchange chromatography (Shah et al., 2010, Encyc. Ind. Biotechnol. Wiley Press; Strauss et al., Biotechnol. Prog., 2010, 26:750). Both follow a similar format QTPP identification → risk analysis → DoE studies → control strategy.

Manageability and cost-benefits need to be considered when implementing QbD and designing DoE studies. The cost of very large DoE studies needs to be weighed against the benefit of the processing flexibility. As the industry moves forward, lab automation and high-through put methods like Q-PCR can tip the equation towards the QbD approach. It must be remembered that there is no single, mandatory approach for QbD, a point stressed by the above mentioned case studies. In fact, many risk tools are available (see the list in ICH Q9). The existing product class knowledgebase can be leveraged for risk assessment purposes and DoE design, especially for products with shared unit operations and where cross-product experience is available like antibodies, hematologic products or cytokines. Finally, it must be remembered that QbD itself is optional, and may not always be the best approach (e.g. very unique or small volume products).

(432a) Taking PAT From the Laboratory to Commercial Manufacturing

Mark A. LaPack[1], Stephen B. Jeffery[2], Steven J. Doherty[3], Joseph R. Martinelli[1] and Christopher L. Burcham[4], (1)CPR&D Process Definition, Eli Lilly and Company, Indianapolis, IN, (2)Process Engineering, Eli Lilly and Company, Kinsale, Ireland, (3)Ms&t PAT, Eli Lilly and Company, Indianapolis, IN, (4)Chemical Product Research and Development; Particle Design Laboratory, Eli Lilly and Company, Indianapolis, IN

Measurements using PAT are applied to two main areas: 1) Gaining process understanding and 2) maintaining process control. The decision to use PAT for control applications is an outcome of the design and development of the process. The parameters to be controlled, the process design space, and control limits dictate the type of technology selected (pH, conductivity, mid-IR, mass spectrometry, etc.) and the user requirements (materials of construction, limit of detection, dynamic range, robustness, etc.).

An example will be presented where a dissolved gaseous byproduct further reacts to form an impurity that affects both product quality and yield. Based on process models, rationales were developed for monitoring either the reaction liquid or the headspace gas during the process designed to remove this gas. The focus of this presentation will be to describe the development of the PAT-based process control system, including discussions about technology selection, vendor selection, risk and contingencies, qualification, installation, and validation.

See more of this Session: Applications of PAT In a Manufacturing Setting
See more of this Group/Topical: Topical I: Comprehensive Quality by Design in Pharmaceutical Development and Manufacture

(432b) Combining Microwave Resonance Technology to Multivariate Data Analysis As a Novel PAT Tool to Improve Process Understanding In Fluid Bed Granulation

Vera Lourenco[1], Thorsten Herdling[2], Gabriele Reich[3], Jose Cardoso Menezes[1], Dirk Lochmann[2] and **Jens Schewitz**[2], (1)Institute for Biotechnology & Bioengineering, I.S.T., Lisbon, Portugal, (2)Quality Operations, PAT, Merck Serono, Darmstadt, Germany, (3)Institut for Pharmazie und Molekulare Biotechnologie, Heidelberg, Germany

A set of 192 fluid bed granulation batches at industrial scale was *in-line* monitored using microwave resonance technology (MRT) to determine moisture, temperature and density of the granules. The gathered data were analysed through multivariate data analysis techniques, such as multiway partial least squares (PLS) and principal component analysis (PCA) and multivariate batch control charts. The combination of all these techniques, along with a few *off-line* particle size measurements, led to a huge gain on the process knowledge and understanding. A seasonality effect could be put into evidence with its influence on the process ongoing and on the final granules particle size. Moreover it was demonstrated and quantified by means of a PLS regression the relation between the particle size and the MRT measurements, highlighting a potential ability of the MRT sensor to predict the final particle size of the granules.

This study contributed successfully to the continuous improvement of an industrial fluid bed granulation process, under the *Process Analytical Technology* (PAT) initiative to meet the *Quality by Design* (QbD) principles.

See more of this Session: Applications of PAT In a Manufacturing Setting
See more of this Group/Topical: Topical I: Comprehensive Quality by Design in Pharmaceutical Development and Manufacture

(432c) PAT Application and Validation of Online Particle Sizing Measurement for Controlling a Critical Quality Attribute

Kevin J. Bittorf, CMC Process & Alliance Management, Strategic Development Networks, Vertex Pharmaceuticals, Cambridge, MA

In accordance with ICH Q8/9/10 and the FDA QbD initiative, companies are adding more and more PAT equipment, installing more sophisticated control systems, and moving to real-time release. Throughout this process, many questions have come to light. How is PAT justified for controlling quality attributes? How is PAT incorporated into the overall control strategy? What is needed for to validate a PAT system? These questions will be answered in a case study of using PAT to monitor and control the critical quality attribute of particle size in a production setting.

See more of this Session: Applications of PAT In a Manufacturing Setting
See more of this Group/Topical: Topical I: Comprehensive Quality by Design in Pharmaceutical Development and Manufacture

(432d) Experimental and Theoretical Study of Lyophilization From a Packing of Vials: Using of Tunable Diode Laser Absorption Spectroscopy (TDLAS) for Process Monitoring and Model Calibration

Pavol Rajniak[1], Cassandra Mifkovich[2] and Leon Farber[2], (1)Pharmaceutical Commercialization Technology, Merck & Co., Inc., West Point, PA, (2)Merck & Co., Inc., West Point, PA

Objectives: 1/experimentally evaluate water sublimation and lyophilization of apharmaceutical formulation from a packing of vials at different conditions;2/monitor total sublimation and desorption rate using a tunable diode laser absorptionspectroscopy (TDLAS) sensor;3/evaluate distribution of heat transfer coefficients from gravimetric watersublimation studies; 4/develop mathematical models for water sublimation andprimary drying from a packing of vials; 5/evaluate cake resistance parametersfrom the primary drying sublimation rates monitored by the TDLAS; 6/perform aqualitative comparison between the SEM images of cakes manufactured atdifferent freezing conditions and cake properties evaluated by the model.

Methods: A simple gravimetric method was used to evaluatethe mass of water sublimed at different locations within a laboratory scale(FTS-Lyostar II) lyophilizer.Unidirectional mathematical model for ice sublimation from a single vial wasused for evaluation of heat transfer coefficients between the shelf and bottomof vials at different locations. Different cumulative distribution functionswere tested for calculation of distribution of evaluated heat transfercoefficients. A mathematical model combining the distribution of heat transfercoefficients with the unidirectional model of primary drying in a vial wasdeveloped and employed for calculation of sublimation rates from a packing ofvials. A simple optimization procedure was developed for evaluation of the cakeresistance parameters (pore size and skin resistance) by fitting the modelpredictions to the TDLAS mass flux data. SEM imaging was used for visualizationof the cake morphology.

Results: Weibull cumulative distribution function accuratelydescribes the distribution of heat transfer coefficients of an inhomogeneousshelf. Mathematical models for both, thepure ice sublimation and sublimation of water from a pharmaceutical(sucrose-based) formulation satisfactorily simulate the sublimation rates measuredexperimentally. A model assuming a homogeneous porous cake was unableto describe experimental observation of a lyophilization cycle withoutannealing. On the other hand, a model assuming the presence of a skin (skinresistance R_{skin} = 0.7) at the cakesurface followed by a porous structure with pore radii 70 microns very well describes experimentalobservations. A faster sublimation was measured using the TDLAS for a cyclewith annealing. The experimental mass flux can be satisfactorily predicted usinga cake model with a thinner skin at the surface (skin resistance R_{skin} = 0.1) followed by a porousstructure with pore radii 90 microns. SEM imaging nicely supports results oftheoretical evaluation showing a thicker skin followed by smaller pores forcakes from the cycle without annealing.

Conclusions: A multi-dimensional mathematical model oflyophilization from packing of vials in combination with the TDLAS probe is auseful tool for better understanding and faster development and scale-up oflyophilization processes.

See more of this Session: Applications of PAT In a Manufacturing Setting
See more of this Group/Topical: Topical I: Comprehensive Quality by Design in Pharmaceutical Development and Manufacture

(432e) Optical Coherence Tomography: A New PAT-Tool for Fast and Non-Destructive Analysis of Tablet Coating Quality

Daniel M. Koller[1], Christoph Schinwald[1], Otto Scheibelhofer[1], Michael Leitner[2] and Johannes G. Khinast[1], (1)Research Center Pharmaceutical Engineering GmbH, Graz, Austria, (2)RECENDT - Research Center for Non Destructive Testing GmbH, Linz, Austria

Optical coherence tomography (OCT) is a non-invasive imaging technique allowing fast and high-quality cross-sectional imaging of semi-transparent and turbid media. Biomedical research and diagnostics have been up to now the main driving forces for the reported applications and progress in OCT. The characteristics of OCT are the ability to provide high-resolution images and depth-resolved information also in strongly scattering media in a contact-free way. Hence OCT is well suited to image layered structures like retina, skin or multi-layered foils in the packaging industry. However, the application of OCT for the analysis and characterization of tablet coatings in the pharmaceutical sciences is still relatively new [1]. The image contrast is due to inhomogeneities in the refractive index of the sample material, and thus, OCT provides complementary information to other high resolution imaging techniques like terahertz imaging, X-ray computed tomography (CT) and magnetic resonance imaging (MRI).

In this work, high-speed spectral domain OCT was demonstrated as a new Process Analytical Process (PAT) tool for the characterization of pharmaceutical tablet coatings [2]. A Design of Experiments (DoE) was created to produce tablet batches with three coating process parameters. The coating thickness was determined via conventional testings with a µm-caliper or weight gain measurements and compared with OCT measurments. In addition, the results were related to investigations carried out with already established PAT tools like Near-Infrared (NIR) and Raman spectroscopy combined with qualitative and quantitative Multivariate Data Analysis (MVDA).

Finally, the influence of process parameters on the dissolution behavior as a critical quality attribute for the tablets was related to the disintegration times of the tablets for response surface modelling.

[1] Mauritz JMA, Morrisby RS, Hutton RS, Legge CH, Kaminski CF 2010. Imaging Pharmaceutical Tablets with Optical Coherence Tomography. Journal of Pharmaceutical Sciences 99(1):385-391.

[2] Koller DM, Hannesschläger G, Leitner M, Khinast JG. Non-Destructive Analysis of Tablet Coatings with Optical Coherence Tomography. European Journal of Pharmaceutical Sciences. 2011.

See more of this Session: Applications of PAT In a Manufacturing Setting

(432f) QbD-Driven PAT Applications for Precision Spray-Coating Development and Control

Charles Miller[1], Bruce T. Thompson[1], Fan Zhang-Plasket[1], Patrick T. Schilling[2], John Higgins[1], Gert Thurau[1], Mano Ramasamy[1] and Eric Ahuja[3], (1)Process Analytical Technologies, Merck and Company, Inc., West Point, PA, (2)Pharmaceutical Commercialization Technology, Merck & Co., Inc., West Point, PA, (3)Director Analytical Chemistry Development and Supply, Merck & Co, Inc, West Point, PA

Two of the core practices of Quality by Design (QbD), design space development and deployment of control strategies, can be greatly facilitated by Process Analytical Technologies (PAT). Ultimately, however, the actual usage of PAT methods for any QbD task must first be warranted by a careful and thorough risk assessment of the product, process and measurement system.

This work addresses one such case, where the product is an API-coated immediate release drug product that is manufactured by a precision spray coating process.

An initial risk assessment of the process indicated that the spray coating operation had considerable risk with respect to two critical quality attributes: content uniformity and composite assay. In response to this risk assessment, a PAT measurement system based on NIR diffuse reflectance spectroscopy was used for two different but equally critical tasks:

1) rapid tablet composition analysis during process development, to assist in design space development and, thus, mitigation of the content uniformity risk, and

2) at-line monitoring of coating growth during the spray coating process, to support the spray coating control strategy, and thus mitigate the composite assay risk

Details regarding the QbD approach, as well as the design and performance of the NIR methods used for development and control, will be discussed. In addition, a brief discussion of the hardware and software infrastructure required to deploy the at-line monitoring PAT method at the supply site will be made.

See more of this Session: Applications of PAT In a Manufacturing Setting
See more of this Group/Topical: Topical I: Comprehensive Quality by Design in Pharmaceutical Development and Manufacture

PROCESS ANALYTICAL TECHNOLOGY FOR RECOMBINANT PANDEMIC FLU VACCINES: VIRAL ULTRASTRUCTURE, AGGREGATION, AND BINDING

De-Hao Tsai,[1] Daniel I. Lipin,[2] Suvajyoti Guha,[1,3] Jeremy I. Feldblyum,[1,3] Kenneth D. Cole,[1] Kurt A. Brorson,[4] Michael R. Zachariah,[1,3] Michael J. Tarlov,[1] Anton P. J. Middelberg,[2] Leonard F. Pease III[1,5]

1. National Institute of Standards and Technology (NIST), Gaithersburg, MD 20899, USA
2. University of Queensland, Brisbane, Australia
3. University of Maryland, College Park, MD 20742, USA
4. Food and Drug Administration, While Oak, MD 10903, USA
5. University of Utah, Salt Lake City, UT 84112, USA

Abstract

Here we describe our progress in developing electrospray differential mobility analysis (ES-DMA) into a process analytical technology (PAT) for recombinant vaccines. The threat of pandemic flu (e.g. avian influenza H5N1, swine flu H1N1, etc.) remains a significant public concern. Recombinantly produced vaccines hold significant advantages over traditional means of growing vaccines including increased production rates, shorter times between viral identification and clinical treatment, and improved safety profiles. However, accelerated production methods also require advanced process control sensors (i.e. process analytical technologies) that can rapidly and accurately detect out-of-specification intermediates. ES-DMA has the potential to fulfill this need. Here we describe our efforts to use ES-DMA to (1) rapidly quantify the assembly state and integrity of recombinantly produced virus like particles (VLPs), (2) determine the viral aggregation state, and (3) quantify antibody-virus binding necessary to evaluate stoichiometry and vaccine efficacy.

Introduction

Recent threats of pandemic flu, including SARS, H5N1 avian influenza and the H1N1 swine flu of 2009, highlight the need for rapid vaccine selection and production.[1] Traditional manufacturing techniques require several months for vaccine stocks to mature, cannot be readily adapted to rapidly changing viral subtypes, and some individuals are allergic to egg based products.[1] Recent recombinant production strategies can overcome each of these challenges.[2-7] In recombinant strategies, cell stocks are induced to produce viral coat proteins that are genetically modified to express antigens specific to the flu virus. The proteins are then subsequently assembled into complete virus like particles (VLPs) before or after purification to remove adventitious (e.g. murine minute virus, Cache Valley virus) and endogenous (e.g. retrovirus) viruses in harmony with US Federal Code.[1,8] Similar recombinant strategies have been used to produce vaccines against the human papillomavirus (HPV), which is marketed by Merek as Gardasil.

To provide additional manufacturing flexibility, the US Food and Drug Administration (FDA) introduced the Quality by Design (QbD) and Process Analytical Technologies (PAT) initiatives. Though novel to many in the chemistry, biology, and biotechnology fields, these initiatives mirror traditional process control efforts well established in Chemical Engineering. Specifically, the PAT initiative calls for additional sensors at key locations within biomanufacturing processes to directly measure critical quality attributes related to the final product composition. Examples of relevance to VLP manufacture include sensors to determine the state of protein assembly, extent of protein and VLP aggregation, and interrogate VLP conjugation. When such sensors are integrated upstream from the final lot release, process control techniques can be used to adjust process settings (if allowed following QbD analysis), abort batches before they undergo very expensive post assembly separation, and to identify why processing goes awry in *post mortem* analysis. In so doing, PATs may increase the net process productivity and decrease the associated manufacturing costs.

Here we present an emerging technique, called electrospray differential mobility analysis (ES-DMA), that holds potential to act as a sensor for each of these applications. In ES-DMA, antibodies, antigens or antibody-antigen complexes in solution are electrosprayed to produce highly charged droplets.[9] These droplets then pass into a charge neutralization chamber, which dries the droplets and fixes the charge on the particles between +2 and -2.[10] Upon entry into the DMA, only positively charged particles are deflected toward a collection slit based on their charge-to-size ratio.[11] Because a vast majority of positively charged particles possess a +1 charge (>95% of 45 nm particles), essentially only particles of a particular size will traverse the collection slit at the exit of the DMA into a condensation particle counter (CPC).[12] Within the CPC, particles pass through a saturated butanol vapor and grow into droplets several microns in size, which can then be counted individually as they obscure light impinging on a photodetector.[9,13,14] The total number of particles passing through the detector is then summed and reported for a particular size. Scanning through a range of sizes assembles a complete size distribution. Peaks in the distribution correspond to individual proteins, protein assemblies (e.g. capsomers), virus & VLPs, or viral aggregates.[8,13-16]

We have previously reported the use of ES-DMA to analyze the size of viruses, VLPs and viral aggregates.[8,13-15] For example, VLPs assembled from solution can be heterogeneous in size or poorly assembled.[14] Figure 1 compares size distributions obtained using ES-DMA to distributions from TEM for VLPs with a 17 amino acid residue from avian influenza.[14] ES-DMA has superior resolution (subnanometer) from 5-500 nm.[17] Similarly, we have recently reported that ES-DMA can be used to rapidly interrogate the composition of the viral coat and that the size measured by ES-DMA most closely matches the edge-to-edge length.[8] We have also shown that temporal and thermal degradation can be tracked using the instrumentation.[8] Here we extend our analyses to viral degradation at low pH and detection of virus-antibody complexes, both of which are important in biomanufacturing environments because they are critical to removing adventitious viral contaminants and evaluating the humoral response induced by vaccines.

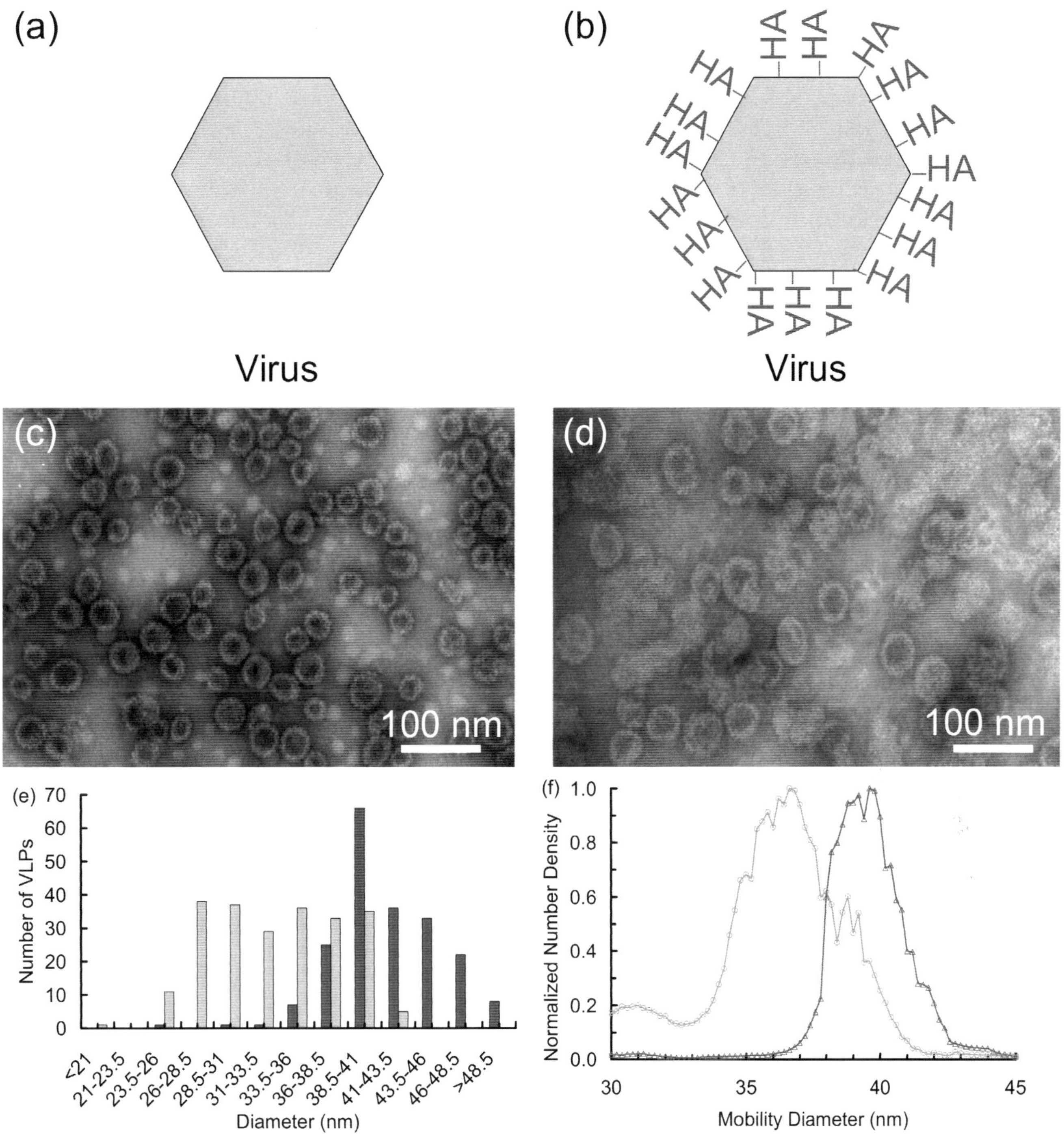

Figure 1. Representation of (a) bare polyoma VLP and (b) polyoma VLP with 17 amino acid reside expressed on external surface loops. TEM micrograph of (c) bare polyoma VLP and (d) polyoma VLP with 17 amino acid reside expressed on external surface loops. Size distributions obtained using (e) TEM and (f) ES-DMA comparing bare polyoma VLP (light bars or circles) and polyoma VLP with 17 aminoacid reside expressed on external surface loops (dark bars or triangles). Adapted from Pease, et al., Biotech. Bioeng., 102 (2009) 845-855.[14]

Materials and Methods

For analysis of acidic degradation, bacteriophage pp7 along with its host *Pseudomonas Aeruginosa* were obtained from the ATCC (Manassas, VA; accession numbers 15692-B4 and 15692). Stocks were prepared by ultracentrifugation using CsCl gradients, and intact phage were enumerated by plaque assays using their respective hosts as described elsewhere[15,18,19] to determine an initial titer of $6.9 \cdot 10^{12}$ pfu/mL. Infectivity loss experiments at low pH were performed by diluting the phage in 10 mmol/L tris at pH 7.5 into 100 mmol/L citric acid/sodium citrate at various pH levels (2.5 to 5.0 and 7.0) for 30 min before pH neutralization by increasing the sample volume by 20% with 3.0 mol/L tris, pH 9.0. Phage in these test articles were enumerated by a plaque assay. Dilutions of the virus stocks and pH treated samples were prepared and added to mid-log phase host and liquefied top agar, which was spread over tryptic soy agar plates. TEM images were obtained as described in detail elsewhere.[8]

For analysis of acidic degradation by ES-DMA, the phage were dialyzed into 10 mmol/L ammonium acetate at pH 7 to remove nonvolatile salts and limit sample conductivity to <0.2 S/m.[14] Studies (not reported) indicate that buffer exchange does not significantly impact phage infectivity. Acidic buffer solutions at pH's of 2.00±0.25, 3.00±0.25, 4.00±0.25 and 5.00±0.25 were prepared by titrating glacial acetic acid into 0.5-1.0 mL of an ammonium acetate solution until the desired pH was measured with an immersed pH probe (#525A, Orion). Ten microliters of the dialyzed virus samples were exposed to 190 μL of these acidified buffer solutions for 30 minutes before quenching the reaction with 190 μL of cold (4°C) ammonium hydroxide. Cold base counters temperature increases associated with mixing of non-ideal acid-base solutions.[20] Because the resulting neutralized test articles could not be electrosprayed due to excessive conductivity, the samples were again dialyzed into an approximately 3.1 mmol/L solution of ammonium acetate at neutral pH. Additional ammonium acetate buffer was added, if necessary, to bring all test article volumes (including controls) to 510±10 μL to achieve uniform levels of dilution.

The bacteriophage MS2 was selected for virus-antibody neutralization assays because it is a well studied biosafety level 1 bacteriophage and known antibodies were available. Phage (#15597-B1 from the ATCC) and E. Coli host (ATCC #15597) were propagated and purified as previously described.[13] MS2 samples were stored at 4°C or frozen at -80°C. A polyclonal antibody labeled P and a monoclonal antibody labeled G were obtained from Tetracore, Inc (Rockville, MD). Antibodies were stored at 4°C until use. The virus and antibodies formed complexes when combined in concentration ratios indicated in the figure captions and allowed to incubate for 4.0 h at ambient temperature before storing at 4°C. To determine the infectivity, incubations were serially diluted and active MS2 virus was measured by plaque assay on the host. Samples at a concentration of $3.45 \cdot 10^{13}$ virus particles/mL but as small as 30 μL were prepared for ES-DMA as follows. Samples were first dialyzed against 2 mmol/L ammonium acetate (neutral pH) using mini dialysis cups (#69590, Pierce, Inc.). Dialysis was necessary to reduce the concentration of nonvolatile salts that may encrust the surface of aerosolized viruses and virus-antibody complexes and to reduce solution conductivities ≤0.2 S/m.[9,14] Arrayed cups were floated on 50 mL of DI water to removed glycerol residues as per the manufacturer's instructions. Approximately 40 μL of sample was then added to each cup and nonvolatile salts allowed to dialyze for at least 3.5 hours at room temperature. The volume of the sample before and after dialysis, which varied from sample to sample, was measured. This dialysis step was previously found to not significantly diminish the infectivity of bacteriophages.[13] Immediately before insertion into the electrospray chamber the samples were diluted 20 fold to extend the amount of sample available for analysis without deleterious effect. Fetal bovine serum was also obtained and diluted in ammonium acetate to the concentration reported in the figure captions.

Solutions were electrosprayed with an Electrospray Aerosol Generator (TSI Inc., USA, #3480), which was operated with gas flow rates of 1.0 L/min of house air and 0.2 L/min of CO_2.[9] A vial containing the sample was placed under pressure, driving the sample solution through a capillary (25 μm in diameter, 24 cm in length) at approximately 66 nL/min. The capillary tip formed a tapered outlet at which voltages up to 3.5 kV were applied as needed until the exiting meniscus formed into a sharp Taylor cone and the measured current remained stable.[9] A stable cone was critical because other meniscus flow regimes fail to produce the narrow distribution of small droplet sizes necessary to achieve high quality, reproducible results. The highly charged droplets, formed at the tip of the Taylor cone, passed into a bipolar charger where the charge remaining on the droplets achieved a steady-state distribution dominated by +1, 0, -1 as described by Wiedensohler.[12] For example, charge neutralization of 45 nm particles (e.g. viruses, proteins, etc.) yields a charge distribution consisting of 61.5% with a net neutral charge, 16.1% with a single positive charge, 0.5% with a double positive charge, 21.1% with a single negative charge, 0.8% with a double negative charge and only 0.01% with charges outside this range.[12] These dry, charged particles (e.g. antibodies, viruses, antibody-virus conjugates) then transited the approximately 1.4 m of plastic Tygon tubing (1.6 cm diameter) connecting the exit of the ES to the entrance of the DMA. Due to adherence of the antibodies to the inside of the electrospray capillaries, quantifying the amount of free antibody before and after complexation to obtain a second supporting measure of antibody binding was not feasible. Adherence was confirmed by replacing the antibody containing sample by 2.0-20 mmol/L ammonium acetate (neutral pH) in the pressure chamber at the end of a run. Rapid sequential size distributions showed at least one peak corresponding to the free antibody, whose magnitude exhibited exponential decreases in consecutive runs as the antibody desorbed from the capillary wall.[21] Within the DMA, samples were separated based on their charge-to-size ratio.[14,22] The DMA (TSI Inc., USA, #3080) was operated with a sheath-to-sample flow rate ratio of 30:1.2 and voltages ranging from 0.0 kV to -10 kV were applied to the central electrode to classify the particles. The flow rate exiting the DMA was 1.0 L/min, which was supplemented by 0.5 L/min of HEPA filtered air as it entered the condensation particle counter (CPC). Within the CPC (TSI Inc., USA, #3025A), particles pass through a saturated butanol environment and grow into droplets several microns in size which can be counted individually as they obscure light impinging on a photodetector. Because a negative bias is applied to the DMA electrode, only positively charged particles are counted by the CPC. Data points were collected at a rate of three per minute with a step interval of 0.2 nm. The raw CPC count is divided by the fraction of positively charged particles described by Wiedensohler[12] to obtain the count for all particles regardless of their charge state when entering the DMA. The DMA voltage is converted into a spherically equivalent size by balancing the mobility of the particle against the mobility of the

DMA tuned by adjusting the voltage and flow rate as described elsewhere.[9,22] Conversion to size was performed assuming the particles to be spheres of diameter d with a Cunningham slip correction factor of $C_c = 1+Kn[\alpha+\beta \exp(-\gamma / Kn)]$, where $Kn = 2\lambda/d$, $\alpha = 1.257$, $\beta = 0.40$, $\gamma = 1.110$, and the gas mean free path at room temperature $\lambda = 66$ nm.[9] The mobility diameter corresponding to the ES-DMA peak maximum is reported as its size. The mobility diameter measured depends on the drag force experienced by the particle, which was proportion to the projected area along each of the particle's principle axes.[8,9,16,22-24]

Results and Discussion

Because we have previously demonstrated ES-DMA as a tool to characterize VLPs as influenza vaccines,[14] the extension of our efforts reported herein focuses on acidic degradation of viruses and antibody neutralization in preparation for future efforts with VLPs and other vaccine constructs.

Low pH Viral Degradation

We now address the degradation in acidic environments of phage pp7, which was chosen as a convenient biosafety level I model for small non-enveloped mammalian viruses.[15,18,19] Like many small viruses, pp7 consists of an RNA genome and an icosahedral capsid assembled from 180 identical copies of a single major capsid protein, having a molecular weight of 13.9 kDa.[15,18,19] (Additional proteins encoded by the pp7 genome probably contribute only marginally to the external capsid structure detected by the DMA due to their significantly lower concentration.) Figure 2 shows the size distributions of pp7 subjected to acid as a function pH. Two sets of peaks are evident, one for intact viruses and one for smaller species. The 23.4 nm peak maximum corresponds to the virus size we have measured previously for pp7.[8,13,15] The smaller peaks (<15 nm) represent free capsid proteins, capsomers, or capsomer fragments. The molecular weight correlation of Bacher, et al.,[10] identifies the primary peak (at 5.4 nm) as a dimer of the major capsid protein. Tars, et al., report that capsids of the viral genus *levivirus* often assemble as covalently linked dimers of the major capsid protein such that *levivirdae* capsids may often be regarded as assemblies of 90 dimers, termed capsomers, instead of 180 individual coat proteins commensurate with our observations.[25]

Several changes in the size distribution take place as pp7 is exposed to acid. First, the concentration of intact virus decreases substantially. Our finding of incomplete capsid degradation at pH 4.0 is in reasonable agreement with the previous work of Brorson, et al.,[26] who find incomplete degradation of the capsid of the xenotropic murine leukemia virus (X-MuLV), another small virus of biomanufacturing interest, at pH 3.8. Second, dimer peaks completely dominate the size distribution. Our data therefore support a mechanism for low pH inactivation where the intact pp7 virus particles disintegrate into dimers without accumulating intermediate species. Thus, ES-DMA can be used as a tool to evaluate virus inactivation mechanisms, which remain essential industry efforts to identify strategies to maximize virus safety.

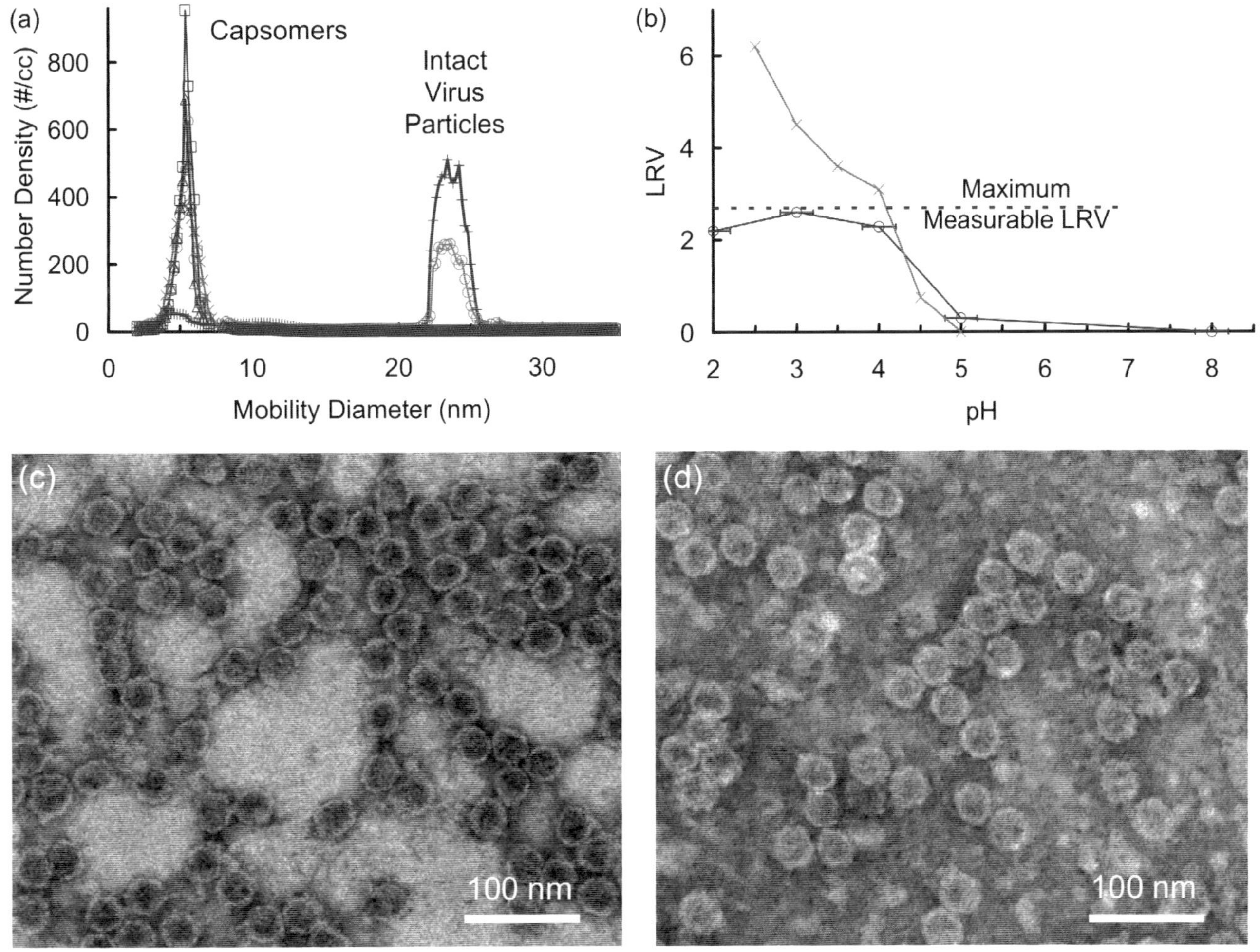

Figure 2. ES-DMA size distributions for (a) pp7 subjected to pH's of neutral (+), 5.0 (o), 4.0 (x), 3.0 (triangle), and 2.0 (square) for 30 minutes. (b) LRV versus pH for pp7 from ES-DMA (o) and plaque assay (x). TEM images of pp7 (c) untreated and (d) exposed to pH ~4.

We can quantify the loss of the intact virus particles using the logs of retained virus (LRV) formalism defined as $LRV=Log_{10}(N_{in}/N_{out})=Log_{10}[(C_{in}V_{in})/(C_{out}V_{out})]$, where N_{in}, C_{in} and V_{in} are the number or moles, concentration, and volume of phage entering a process stream or at the beginning of a batch degradation process and N_{out}, C_{out} and V_{out} are the number or moles, concentration, and volume exiting the stream or at the termination a batch process.[8,27] The LRV represents a convenient means of representing concentration changes that can span several orders of magnitude in a unit operation. Larger LRV's represent greater viral clearance via removal or inactivation (i.e. a concentration decreases by three orders of magnitude from entrance to exit is described as changing by 3 $\log_{10}$ or an LRV of 3). Here we approximate the concentration ratio as the ratio of the counts under the peak (which ranges from 21.8 nm to 26.2 nm) of the control at neutral pH to the counts at the pH of interest. Figure 2b illustrates how the LRV measured by ES-DMA and plaque assays depends on pH. We find minimal changes in the LRV in the neutral range between pH 5 and 8, a large jump between pH 4 and 5, and a plateau below pH 4 for ES-DMA but not the plaque assay. This is because the plaque assay has a large linear range (up to 8 $\log_{10}$ in this experiment), whereas the dynamic range of ES-DMA is limited by its lower limit of detection (≤10^9 virus particles/mL).[27,28] Although ES-DMA boasts single particle counting like the plaque assay, the 66 nL/min flow through the electrospray is effectively diluted into a gas flow of 1.5 L/min, decreasing the apparent concentration by $\approx 2.3 \cdot 10^7$. Thus, a sample that begins at $6 \cdot 10^{11}$ viruses/mL, for example, is diluted by a factor of 50 in the preparation protocol, and then is further diluted in the instrument by $2.3 \cdot 10^7$ would be expected to sustain a maximum gas-phase count of 500 particles/cubic centimeter (see Figure 2a). Because the baseline has a number density of approximately unity, the maximum LRV measurable with this protocol and sample set is $\log_{10}(500/1) = 2.7$. This example implies that the plateau ≤pH 4 in Figure 2b is due to limitations of the measurement protocol and not degradation of the sample. The presence of the plateau indicates the uncertainty in these measurements for which one standard deviation at pH 2-4 equals 0.21 in $\log_{10}$ units. This uncertainty is comparable to that of the plaque assay, which has an uncertainty of approximately 0.3 $\log_{10}$ units.[27] We note in contrast that our method does not require long incubation times nor specific hosts. Indeed, the technique is fully functional even for viruses for which neither the genetic sequence nor the host has been fully characterized. Thus, within their respective dynamic ranges, curves derived from ES-DMA and the plaque assay essentially overlap within the linear range of the former. This argues that pp7 inactivation can be completely attributed to disintegration of the capsid into capsomers, rather than a more subtle effect such as denaturation of surface coat proteins.

Viral Neutralization

ES-DMA assays of nanoconjugate structure and stoichiometry rely on changes in the measured mobility size upon binding (see Fig. 3a).[9,24,29] Here we report on the virus-antibody conjugates formed by binding anti-MS2 antibodies (i.e. those specific to the bacteriophage MS2) to MS2. As antibodies recognize and bind to the viral coat, the antibody-virus complex forms, increasing the apparent size of the virus in the size distributions. Figure 3 shows that prior to binding the MS2 virus particle displays a symmetric monomodal peak at 23.2 nm in reasonable agreement with previous results of Cole, et al.,[13] and Hogan, et al.,[28] who find MS2 to be 23.0 nm and 24.1 nm, respectively. After binding Figure 3b the apparent size of the complex increases with larger solution concentrations of antibodies leading to further increases in size. Both the polyclonal P and monoclonal G antibodies increase the apparent size of the virus peak, though changes in size due to the former are more modest than those due to the latter. The apparent size of the virus complex as a function of the antibody concentration is summarized in Figure 3d using a dose response curve. Although some of the 2.2 nm shift may be due to one or more free antibodies clustering with the complex formed in solution as the droplet dried,[16,21,30] quantitative analysis (not shown) indicates that nonspecific aggregation alone cannot explain the size shifts observed, particularly for the G antibody, demonstrating that the ES-DMA cleanly distinguish antibodies that bind well from those that do not. We note in passing that these results support the MS2 specificity of both the P and G antibodies because nonspecific human antibodies mixed with MS2 showed no indication of complex formation in ES-DMA (data not shown). These results demonstrate the utility of size based discrimination of virus-antibody binding.

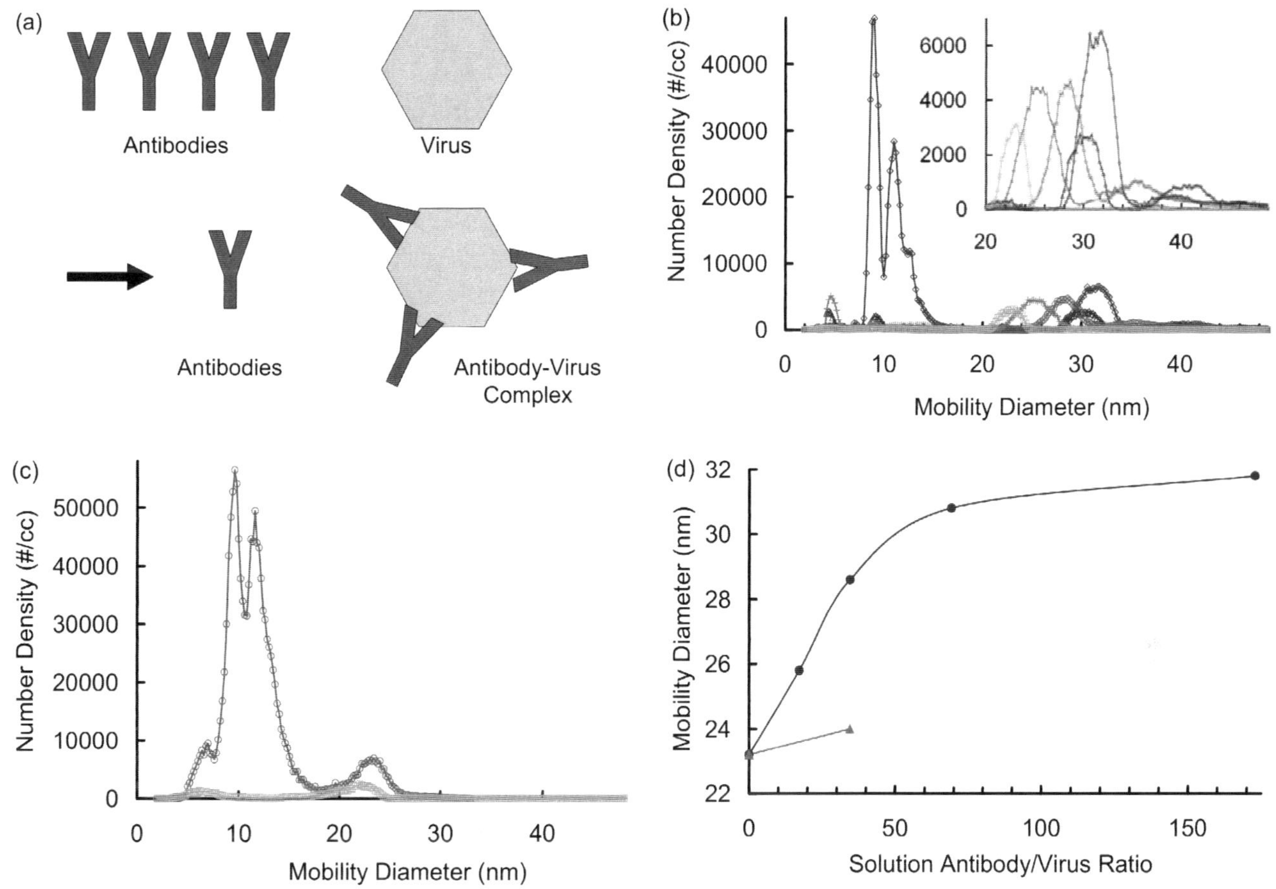

Figure 3. (a) Reaction scheme for antibody-antigen (viral) binding suggesting that the apparent size of the virus-antibody complex should exceed that of the bare virus. (b) ES-DMA size distributions of MS2 virus (squares) with monoclonal antibodies at molar solution concentration ratios of 173 (diamond), 69 (triangle), 35 (circle), and 17 (+) antibodies per virus. (c) ES-DMA size distributions of MS2 virus (squares) with 35 (circle) polyclonal antibodies/virus. (d) Summary dose-response curves of mobility size from ES-DMA versus solution antibodies-virus ratio for monoclonal (circle) and polyclonal antibodies (triangle).

A remarkable feature of the ES-DMA data for the G antibody is the presence of two peaks for the antibody-virus complexes as seen in the inset of Figure 3b. For example, at the highest G antibody concentration, peaks are observed at both 31 nm corresponding to one antibody bound virus and 40 nm corresponding to two antibody labeled virus particles. Similar bimodal distributions are observed at other concentrations. The initial virus sample (Fig. 3b) did not show any signs of virus dimerization, indicating the antibody to be responsible for formation of the two-virus complexes. We suspect that the antibody serves as a linker for the two viruses because each antibody possess two Fc regions, both of which can bind. The presence of bimodal distributions for the G antibody and their absence for the P antibody provides further evidence of the binding strength of the former. This result is particularly important because viral aggregation is one of the three primary ways in which an antibody can neutralize virus; ES-DMA holds the potential to rapidly and clearly distinguish this form of neutralization.

Although it is clear that increases in the apparent size of the virus peak reflect antibody binding, the exact correlation between size and stoichiometry remains uncertain. Two models are available to analyze the data, both of which approximate both antibody and viruses as spheres of diameter d_{ab} and d_v, respectively. Laschober, et al.,[29] assert based on volume conservation that the number of antibodies, n_{ab}, bound to a virus particle is equal to the volume of the thin spherical shell, $\pi(d_{av}^3-d_v^3)/6$, divided by the volume of an individual antibody, $\pi d_{ab}^3/6$,

$$n_{ab} = \frac{d_{av}^{3} - d_{v}^{3}}{d_{ab}^{3}} \quad . \tag{1}$$

A similar expression was used by Pease, et al., to approximate the surface binding of thiol terminated single stranded DNA on gold nanoparticles.[9] A second model more recently advanced by Pease, et al., draws upon a fundamental analysis of the forces acting on the complex within the DMA.[8,9,13-17,22-24,27,28,30-34] Briefly, the drag force acting on the complex within the DMA is proportional to the projected area of that complex when it is smaller than the mean free path of the carrier gas. By definition each complex has three principle axes.[22,35] A Brownian average of the projected area, A_i, orthogonal to each of these axes finds

$$d = \left[\frac{\sqrt{\pi}}{6} \sum_{i=1}^{3} A_i^{-\frac{1}{2}} \right]^{-1} \tag{2}$$

in the free molecular regime.[9,22] This equation has been successfully used in several recent articles to approximate the surface density of DNA on nanoparticles, size virus clusters, determine the number of quantum dots attached to bacteriophage λ, and evaluate the number of streptavidins on rod shaped nanoparticles, *inter alia*.[8,9,13-17,22-24,27,30-34]

Table 1 summarizes the predictions of the two equations for the MS2 data reported herein and data for human rhinovirus (HRV) serotype 2, as reported by Laschober, et al.[29] We include the latter with pause because the corresponding size distributions contain several peaks that were not clearly identified, perhaps due to degraded or impure virus samples. The uncertainty in stoichiometry using Eq. 1 based on an uncertainty in complex mobility diameter of ±0.5 nm is approximately ±4 antibodies as indicated by Laschober, et al.,[29] while the uncertainty using Eq. 2 remains less than ±1 antibody. In either case, the disagreement between the two models is significant. Laschober, et al., support their estimates by asserting that their construction is intuitively simple and that their predicted number of antibodies is approximately equal to the close packed limit of spherical antibodies on the viral surface in agreement with other estimates for HRV.[29] Proponents of the second model assert that it is more rigorous than the former because it is based on the fundamental physics underlying the measure of the complex's mobility within the DMA, the second equation can be derived from first principles whereas the first cannot, and the close packed limit is unlikely to be achieved in practice due to steric and electrostatic repulsion between neighboring antibodies.[11,22,36] Both models have been used previously in the literature and both make unphysical simplifying assumptions about the structure antibodies by asserting that they are spheres when it is well known that they have a Y-like structure. Final arbitration between the two models awaits more rigorous and advanced computational models—outside of the scope of this contribution—that account for the Y-like antibody structure, antibody orientation relative to icosahedral viral surface, intermolecular forces acting between antibodies and the virus particle as well as between neighboring antibodies, and advanced drag force calculations in the free molecular regime.

Table 1. Stoichiometry of Antibody-Virus Complexes

Sample (solution antibody:virus molar ratio)	Virus Size (nm)	Antibody Size (nm)	Antibody-Virus Complex Size (nm)	Complex Stoichiometry from Eq. 2	Complex Stoichiometry from Eq. 1
MS2 + G (17:1)	23.2	9.0	25.8	1.8	6.4
MS2 + G (35:1)	23.2	9.0	28.6	3.8	15.0
MS2 + G (69:1)	23.2	9.0	30.8	5.4	23.0
MS2 + G (173:1)	23.2	9.0	31.8	6.1	27.0
MS2 + P (35:1)	23.2	9.0	24.0	0.6	1.8
HRV2 + mAB-8F5 (30:1)	29.8	8.8	34.3	4.7	20.4
HRV2 + mAB-8F5 (60:1)	29.8	8.8	35.6	6.1	27.4
HRV2 + mAB-8F5 (90:1)	29.8	8.8	35.6	6.1	27.4
HRV2 + mAB-8F5 (150:1)	29.8	8.8	35.6	6.1	27.4

Binding of antibodies to the virus surface also neutralizes viral infectivity. To test the ability of the polyclonal P and monoclonal G antibodies to neutralize the infectivity of MS2, antibody-virus complexes were introduced to their *E. Coli* host. Fig. 4a shows both the mono and polyclonal antibodies substantially decrease the infectivity. As expected the G antibody was much more effective at decreasing the viral infectivity, to as little as 0.5% at the highest included concentration, than the polyclonal P due to its better binding stoichiometry. For both antibodies, increasing the solution concentration of the antibody leads to more effective neutralization in reasonable agreement with the dose-response curves above. Combining these results with stoichiometry determined by either equation shows that all reductions in infectivity fall on a single power-law fit. The uncertainty in the plaque assay remains commensurate with that above (though log-linear plots tend to emphasize uncertainty at smaller remaining infectivities). *This remarkable result indicates that decrements in activity or infectivity are primarily a function of binding stoichiometry and not antibody identity.* The infectivity decreases by one order of magnitude for every 3.2 or 13 antibodies bound, using Eq. 2 and 1, respectively. Furthermore, these results indicate that just four antibodies can decrease infectivity by 95% and seven antibodies decrease infectivity by 99% using Eq. 2. This implies that less than 8% of the 90 MS2 capsomers must be directly bound to achieve a 99% reduction in infectivity. In contrast, Eq. 1 requires 17 and 31 for 95% and 99% infectivity reduction, respectively. The dramatic difference in predicted neutralizing power emphasizes the need for more comprehensive modeling of binding and gas phase mobility.

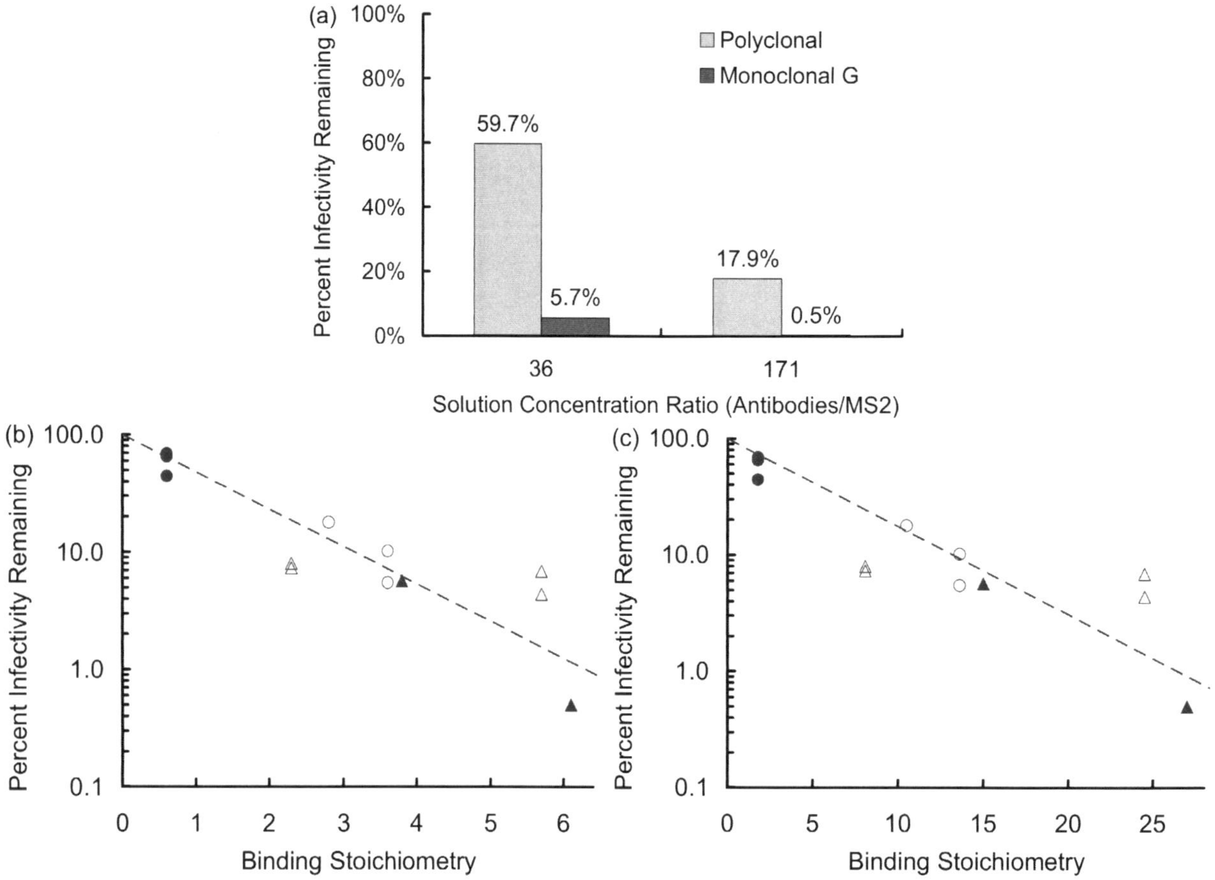

Figure 4. Percent infectivity remaining after antibody binding versus (a) solution concentration of neutralizing anti-MS2 polyclonal P and monoclonal G antibodies. Percent infectivity remaining, *PIR*, for both P (circles) and G (triangles) antibodies versus binding stoichiometry, n_{ab}, as determined by (b) Eq. 2 and (c) by Eq. 1 with close and open symbols representing directly measured and curvilinearly interpolated/extrapolated stoichiometries (from Figure 3d), respectively. The dashed line represents best fits including the point (0, 100%) of $PIR=10^{(2-0.315\ nab)}$ with R^2 = 0.78 for Eq. 2 and $PIR=10^{(2-0.075\ nab)}$ with R^2 = 0.74 for Eq. 1.

Finally, many neutralization assays do not take place in clean solutions containing only viruses and antibodies in nonvolatile ammonium acetate solutions, but occur in more challenging solutions and suspensions including patient serum. Figure 5 provides a size distribution for fetal bovine serum (fbs). Without dilution, the size distribution only increases in the range of 1-50 nm, but with increasing dilution, the peaks tail at progressively smaller sizes. With a twenty fold dilution, the serum background decays by 30-35 nm, allowing larger viruses to be evaluated using this neutralization assay. With 2000 fold dilution, the serum background decays by 12-13 nm allowing even smaller viruses such as pp7 to be analyzed using this assay. At this level of dilution, the large aggregates represented in Figure 5 are no longer formed but individual peaks for the albumins and globulins are observed.

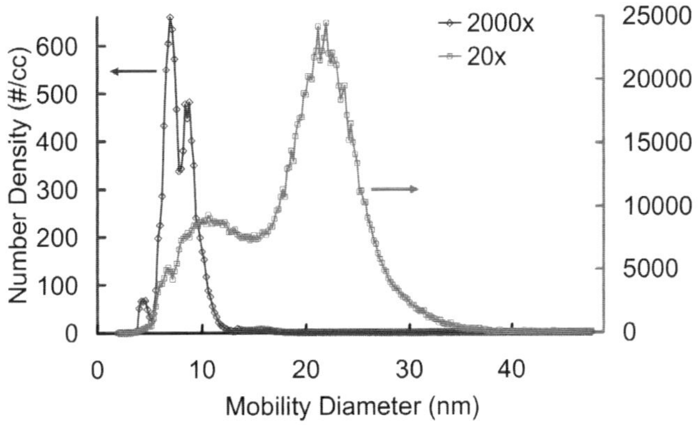

Figure 5. ES-DMA size distribution of positively charged molecular ions from fetal bovine serum diluted twenty (squares) and 2000 (diamonds) times in ammonium acetate buffer.

Conclusion

In summary, ES-DMA represents a powerful new technique to evaluate virus, VLP and vaccine properties and determine the relationship between their structure and properties. The technique is rapid, quantitative, and holds potential to quantitatively analyze binding stoichiometry and kinetics. Because of its design, one particle gives a single count regardless of its mass, enabling precise measurements even at the smallest sizes. In contrast to traditional biological assays, this method does not require long incubation times nor specific hosts. Indeed, the technique is fully functional even for viruses for which neither the genetic sequence nor the host has been fully characterized. These attributes make ES-DMA an attractive candidate with which to evaluate virus neutralization mechanisms. A highlight of the analysis above is that decrements in infectivity are primarily a function of antibody binding stoichiometry and not antibody identity. Future work will apply this technique to characterization of VLPs and adaptation of this technique to at-line analysis in biomanufacturing facilities.

Acknowledgements and Disclaimer

Reference to commercial equipment, supplies, or software neither implies its endorsement by the National Institute of Standards and Technology (NIST) or the Food and Drug Administration (FDA) nor that it is necessarily best suited for this purpose. The authors express appreciation to Tetracore (Rockville, MD) for their generous provision of antibodies specific to MS2. LFP recognizes provision of University of Utah startup funds supporting this endeavor.

References

1. Middelberg APJ, Rivera-Hernandez T, Wibowo N, et al. A microbial platform for rapid and low-cost virus-like particle and capsomere vaccines. *Vaccine.* 2011:doi:10.1016/j.vaccine.2011.1005.1075.

2. Hanslip SJ, Zaccai NR, Middelberg AP, Falconer RJ. Intrinsic fluorescence as an analytical probe of virus-like particle assembly and maturation. *Biochem Biophys Res Commun.* Oct 24 2008;375(3):351-355.
3. Hanslip SJ, Zaccai NR, Middelberg AP, Falconer RJ. Assembly of human papillomavirus type-16 virus-like particles: multifactorial study of assembly and competing aggregation. *Biotechnol Prog.* Mar-Apr 2006;22(2):554-560.
4. Pattenden LK, Middelberg AP, Niebert M, Lipin DI. Towards the preparative and large-scale precision manufacture of virus-like particles. *Trends Biotechnol.* Oct 2005;23(10):523-529.
5. Middelberg AP, Rivera-Hernandez T, Wibowo N, et al. A microbial platform for rapid and low-cost virus-like particle and capsomere vaccines. *Vaccine.* Jun 7 2011.
6. Liew MW, Rajendran A, Middelberg AP. Microbial production of virus-like particle vaccine protein at gram-per-litre levels. *J Biotechnol.* Oct 15 2010;150(2):224-231.
7. Ding Y, Chuan YP, He L, Middelberg AP. Modeling the competition between aggregation and self-assembly during virus-like particle processing. *Biotechnol Bioeng.* Oct 15 2010;107(3):550-560.
8. Pease LF, Tsai DH, Brorson KA, Guha S, Zachariah MR, Tarlov MJ. Physical characterization of icosahedral virus ultra structure, stability, and integrity using electrospray differential mobility analysis. *Anal Chem.* Feb 8 2011;83:1753-1759.
9. Pease LF, Tsai DH, Zangmeister RA, Zachariah MR, Tarlov MJ. Quantifying the surface coverage of conjugate molecules on functionalized nanoparticles. *J Phys Chem C.* Nov 22 2007;111(46):17155-17157.
10. Bacher G, Szymanski WW, Kaufman SL, Zollner P, Blaas D, Allmaier G. Charge-reduced nano electrospray ionization combined with differential mobility analysis of peptides, proteins, glycoproteins, noncovalent protein complexes and viruses. *J Mass Spectrom.* Sep 2001;36(9):1038-1052.
11. Knutson EO, Whitby KT. Aerosol classification by electrical mobility: apparatus, theory, and applications. *J. Aerosol. Sci.* 1975;6:443-451.
12. Wiedensohler A. An approximation of the bipolar charge-distribution for particles in the sub-micron size range. *J. Aerosol Sci.* Jun 1988;19(3):387-389.
13. Cole KD, Pease LF, Tsai DH, et al. Particle concentration measurement of virus samples using electrospray differential mobility analysis and quantitative amino acid analysis. *J Chromatogr A.* Jul 24 2009;1216(30):5715-5722.
14. Pease LF, Lipin DI, Tsai DH, et al. Quantitative characterization of virus-like particles by asymmetrical flow field flow fractionation, electrospray differential mobility analysis, and transmission electron microscopy. *Biotechnol Bioeng.* Feb 15 2009;102(3):845-855.
15. Lute S, Riordan W, Pease LF, et al. A consensus rating method for small virus-retentive filters. I. Method development. *PDA J Pharm Sci Technol.* Sep-Oct 2008;62(5):318-333.
16. Pease LF, Elliott JT, Tsai DH, Zachariah MR, Tarlov MJ. Determination of protein aggregation with differential mobility analysis: application to IgG antibody. *Biotechnol Bioeng.* Dec 15 2008;101(6):1214-1222.
17. Tsai DH, Zangmeister RA, Pease LF, Tarlov MJ, Zachariah MR. Gas-phase ion-mobility characterization of SAM-functionalized Au nanoparticles. *Langmuir.* Aug 19 2008;24(16):8483-8490.

18. Lute S, Aranha H, Tremblay D, et al. Characterization of coliphage PR772 and evaluation of its use for virus filter performance testing. *Appl Environ Microbiol.* Aug 2004;70(8):4864-4871.
19. Lute S, Bailey M, Combs J, Sukumar M, Brorson K. Phage passage after extended processing in small-virus-retentive filters. *Biotechnol Appl Biochem.* Jul 2007;47(Pt 3):141-151.
20. Smith JM, VanNess HS, Abbott MM. *Introduction to Chemical Engineering Thermodynamics.* 5th ed. New York: McGraw-Hill; 1996.
21. Kaufman SL, Skogen JW, Dorman FD, Zarrin F, Lewis KC. Macromolecule analysis based on electrophoretic mobility in air: globular proteins. *Anal Chem.* Jun 1 1996;68(11):1895-1904.
22. Pease LF, Tsai DH, Hertz JL, Zangmeister RA, Zachariah MR, Tarlov MJ. Packing and size determination of colloidal nanoclusters. *Langmuir.* Jul 6 2010;26(13):11384-11390.
23. Pease LF, Sorci M, Guha S, et al. Probing the nucleus model for oligomer formation during insulin amyloid fibrillogenesis. *Biophys J.* Dec 15 2010;99(12):3979-3985.
24. Pease LF, Feldblyum JI, Depaoli Lacaerda SH, et al. Structural analysis of soft multicomponent nanoparticle clusters. *ACS Nano.* Nov 23 2010;4(11):6982-6988.
25. Tars K, Fridborg K, Bundule M, Liljas L. The three-dimensional structure of bacteriophage PP7 from Pseudomonas aeruginosa at 3.7-A resolution. *Virology.* Jul 5 2000;272(2):331-337.
26. Brorson K, Krejci S, Lee K, Hamilton E, Stein K, Xu Y. Bracketed generic inactivation of rodent retroviruses by low pH treatment for monoclonal antibodies and recombinant proteins. *Biotechnol Bioeng.* May 5 2003;82(3):321-329.
27. Guha S, Pease LF, Brorson KA, Tarlov MJ, Zachariah MR. Evaluation of electrospray differential mobility analysis for virus particle analysis: potential applications for biomanufacturing. *J. Virolog. Meth.* 2011:submitted.
28. Hogan CJ, Kettleson EM, Ramaswami B, Chen DR, Biswas P. Charge reduced electrospray size spectrometry of mega- and gigadalton complexes: Whole viruses and virus fragments. *Anal Chem.* Feb 1 2006;78(3):844-852.
29. Laschober C, Wruss J, Blaas D, Szymanski WW, Allmaier G. Gas-phase electrophoretic molecular mobility analysis of size and stoichiometry of complexes of a common cold virus with antibody and soluble receptor molecules. *Anal Chem.* Mar 15 2008;80(6):2261-2264.
30. Pease LF. Optimizing the yield and selectivity of high purity nanoparticle clusters. *J. Nanopart. Res.* 2011;13 2157-2172.
31. Anumolu R, Gustafson JA, Magda JJ, Cappello J, Ghandehari H, Pease LF. Fabrication of highly uniform nanoparticles from recombinant silk-elastin-like protein polymers for therapeutic agent delivery. *ACS Nano.* Jul 5 2011;5:5374-5382.
32. Pease LF, Tsai DH, Fagan JA, et al. Length distribution of single-walled carbon nanotubes in aqueous suspension measured by electrospray differential mobility analysis. *Small.* Dec 2009;5(24):2894-2901.
33. Tsai DH, Pease LF, Zangmeister RA, Tarlov MJ, Zachariah MR. Aggregation kinetics of colloidal particles measured by gas-phase differential mobility analysis. *Langmuir.* Jan 6 2009;25(1):140-146.

34. Yim PB, Clarke ML, McKinstry M, et al. Quantitative characterization of quantum dot-labeled lambda phage for Escherichia coli detection. *Biotechnol Bioeng.* Dec 15 2009;104(6):1059-1067.

35. Happel J, Brenner H. *Low Reynolds number hydrodynamics*. The Hague: Kluwer Academic Publishers 1983.

36. Russel WB, Saville DA, Schowalter WR. *Colloidal Dispersons*. Cambridge, UK: Cambridge University Press; 1989.

(497a) Transferring Monitoring Models Between Different Scales Through Multivariate Statistical Techniques

Emanuele Tomba[1], Pierantonio Facco[1], Fabrizio Bezzo[1], Salvador Garca-Muoz[2] and Massimiliano Barolo[1], (1)DIPIC - Department of Chemical Engineering Principles and Practice, University of Padova, Padova, Italy, (2)Process Modeling & Engineering Technology, Pfizer Worldwide Research & Development, Groton, CT

The challenge of transferring a process monitoring model across different scales is a different problem than the one addressed in previous studies that have focused on transferring operating procedures looking to produce the same grade of product [1, 2, 3, 4].

In the development of a new pharmaceutical product, it is of paramount importance to shorten the launch time of the product, and for new compounds it may be that this is the first time the product is manufactured at pilot scale. At the pilot scale, an extended experimental campaign can be carried out to gain process understanding and disclose potential pitfalls in the process equipment, operating conditions or control configuration. This experimental campaign eventually leads to the definition of a process equipment layout as well as a set of proper ("normal") process operating conditions that can guarantee the required product quality with acceptable variability. If the experimental work at the pilot scale is extended to produce a data set where common cause variability can be observed, then a multivariate statistical process monitoring model can be designed and used to detect (and possibly diagnose) process faults and malfunctions at the pilot-plant scale in real time, thus gaining further process understanding.

When the product is transferred to the commercial manufacturing scale, it is highly desired to have an effective process monitoring framework available as quickly as possible. Carrying out an extended experimental campaign to produce data (where the common cause variability can be observed) is not a viable option at this scale, and therefore the issue arises on whether the monitoring model developed at the pilot-plant scale can be transferred up to the production scale to monitor the manufacturing process until a sufficient amount of data have been collected to design a full production-scale monitoring model.

This work addresses the above issue. Some approaches will be presented that use multivariate statistical techniques [1, 5, 6] and enable transferring a process monitoring model between different scales. The approaches are adaptive, in that initially they rely heavily on the wealth of information embedded in the pilot-scale data, but the relevance of these data is progressively discounted as long as more production-scale data become available. The proposed approaches are tested and critically evaluated using real-world data coming from a spray-drying process [7], and prove to be very effective at monitoring normal conditions and quickly detecting process faults at the production scale using data from the pilot scale. The presented methods will accelerate the implementation of continuous quality assurance programs in pharmaceutical manufacturing by transferring a greater portion of the knowledge generated at the pilot scale.

References

[1] Garcìa-Muñoz, S., J. F. MacGregor, and T. Kourti (2005). Product transfer between sites using Joint-Y PLS. *Chemom. Intell. Lab. Sys.*, **79**, 101-114.

[2] Jaeckle, C. M., and J. F. MacGregor (2000). product transfer between plants using historical process data. *AIChE J.*, **46**, 1989-1997.

[3] Jaeckle, C. M., and J. F. MacGregor (2000). Industrial application of product design through the inversion of latent variable models. *Chemom. Intell. Lab. Sys.*, **50**, 199-210

[4] Jaeckle, C. M., and J. F. MacGregor (1998). Product design through multivariate statistical analysis of process data. *AIChE J.*, **44**, 1105-1118.

[5] Jackson, J. E. (1991). A user's guide to principal components. John Wiley & Sons Inc., New York (U.S.A.).

[6] Geladi, P. and R. Kowalski (1986). Partial least squares regression: a tutorial. *Anal. Chim. Acta*, **185**, 1-17.

[7] Dobry E.D., D.M. Settell, J.M. Baumann, R.J. Ray, L.J. Graham and R.A. Beyerinck (2009). A model-based methodology for spray-drying process development. *J. Pharm. Innov.*, **4**, 133-142.

See more of this Session: Applications of Multivariate Data Analysis In Pharmaceutical Process Development: Traditional and Advanced Multivariate Methods
See more of this Group/Topical: Topical I: Comprehensive Quality by Design in Pharmaceutical Development and Manufacture

(497b) Implementation of Genetic Algorithms In the Generation of High-Order Statistical Models

Brendan C. Mack, Nathan Domagalski, Daniel Hallow, Michael Fenster, Lindsay Hobson and Jose Tabora, Bristol-Myers Squibb Company, New Brunswick, NJ

To support the quality by design (QbD) philosophy, it is critical to understand the behavior of chemical reactions used for API manufacturing. The parameter response space may be mapped experimentally and further explored by mechanistic or statistical modeling. In this work, the formation of a pharmaceutical intermediate was studied via a series of fraction factorial experimental designs. The results were then used to construct statistical models for several responses, including starting material conversion, desired product formation, and key impurity formation. The models were constructed using a genetic algorithm to select optimal variables from all the possible linear, collinear and quadratic terms. Although the models demonstrated good agreement with many of the experimental responses, product formation was not accurately predicted suggesting either experimental error of inadequate representation from a quadratic statistical model. To improve the models different approaches were explored to increase the order of the statistical model. The resulting models had much better predictive power than the quadratic models. This talk will discuss the systematic construction of high order statistical models and their implementation in the QbD paradigm.

See more of this Session: Applications of Multivariate Data Analysis In Pharmaceutical Process Development: Traditional and Advanced Multivariate Methods
See more of this Group/Topical: Topical I: Comprehensive Quality by Design in Pharmaceutical Development and Manufacture

(497c) Understanding the Impact of the Properties of Raw Materials Onto Product Quality Using JR-PLS

Salvador Garca-Muoz, Process Modeling & Engineering Technology, Pfizer Worldwide Research & Development, Groton, CT

Multiple proposals are found in literature to analyze the impact of raw materials onto product using data analysis. From the methods proposed, the L-PLS is the only method that not only considers the ratio in which ingredients are mixed, but also the physical properties of the lots of raw material. This method however is limited to analyze data from materials of the same nature such that all the physical properties apply to all materials. This is not the case when a product includes materials that need to undergo different physcial/chemical tests due to theiir nature. Modeing this scenario using the L-PLS would include blocks of missing data with non-overlapping rows or columns.

A new method is presented to handle this situations, the Joint-R PLS (JRPLS) combines the JYPLS and the LPLS methods providing a plattform to model the impact of the physical/chemical properties of multiple ingredients of different natures onto the final quality attributes of the product. The use and advantage of the model is shown in the troubleshooting of multiple scenarios where a complex relation of physical/chemical properties across multiple ingredients was identified as the root-cause of a situation being investigated.

See more of this Session: Applications of Multivariate Data Analysis In Pharmaceutical Process Development: Traditional and Advanced Multivariate Methods
See more of this Group/Topical: Topical I: Comprehensive Quality by Design in Pharmaceutical Development and Manufacture

(497d) De-Risking Scale-up of a High Shear Wet Granulation Process Using Latent Variable Modeling and near Infrared Spectroscopy

Koji Muteki, Ken Yamamoto, George L. Reid and Mahesh K. Krishnan, Pfizer Inc, Groton, CT

In the development of wet granulated drug products, two primary sources of variance (disturbance) include the operational scale of the high shear wet granulation (HSWG) process and Active Pharmaceutical Ingredient (API) lot-to-lot variability, particularly for formulations containing a high drug load. This paper presents a novel Process Analytical Technology (PAT) strategy using Latent Variable Modeling (LVM) with Near Infrared (NIR) spectroscopy to reduce risk in scale-up operations of the HSWG process while simultaneously accounting for API lot-to-lot variability, even with limited manufacturing history. The process involves building a Partial Least Square (PLS) model among the API material properties, the HSWG process parameters and the NIR end points of the HSWG process based on small scale DOE (Design of Experiment) batches. The PLS model is then used in an optimization framework to find suitable small scale mechanical process parameters (impeller/chopper speed) that approximate a previous large scale operation so as to keep the NIR end point of large scale operation constant. Prior to making additional large scale batches with a new lot of API, NIR end points of large-scale HSWG with the new API lot are predicted based on the PLS model developed from the small-scale operation. If the predicted NIR end point for the HSWG using the new API lot is not within the target region, the risks associated with the scale-up operation can then be significantly reduced by modifying other HSWG process parameters such as the total amount of water added or total granulation time to achieve the target region. A case study is presented that demonstrates the effectiveness of this methodology during development and scale-up of a drug product manufactured using a HSWG process.

See more of this Session: Applications of Multivariate Data Analysis In Pharmaceutical Process Development: Traditional and Advanced Multivariate Methods
See more of this Group/Topical: Topical I: Comprehensive Quality by Design in Pharmaceutical Development and Manufacture

(497e) Diverse Uses of Latent Variable Methods In Vaccines and Biologics Manufacturing

Charles Miller, Louis Obando and John Higgins, Manufacturing Division, Process Analytical Technologies (PAT), Merck and Company, Inc., West Point, PA

It is generally understood that latent variable multivariate modeling methods have great value-add potential in vaccines and biologics enterprises. What is relatively less-known, however, is the diversity of tasks that these tools can support in such applications. Although the fundamental math and statistics behind these tools are fixed, each of these different tasks require unique considerations with respect to several attributes, including data alignment/integration, data preprocessing and scaling, model validation, computing hardware, and visualization of results.

This paper will use recent case studies to illustrate the various ways that latent variable methods support current vaccines and biologics efforts at Merck, which include QbD initiatives. These case studies will include retrospective analyses of existing operations for troubleshooting, exploratory modeling of processes for continuous improvement efforts, multi-scale models for scale-down model qualification, and consistency monitoring of manufacturing operations.

See more of this Session: Applications of Multivariate Data Analysis In Pharmaceutical Process Development: Traditional and Advanced Multivariate Methods
See more of this Group/Topical: Topical I: Comprehensive Quality by Design in Pharmaceutical Development and Manufacture

(497f) Insights Into Lactate Metabolism Through Multivariate Analysis of Cell Culture Bioprocess Data

Huong Le[1], Santosh Kabbur[2], Ziran Sun[3], Luciano Pollastrini[3], Kevin Johnson[3], Keri Mills[3], George Karypis[2] and Wei-Shou Hu[1], (1)Chemical Engineering and Materials Science, University of Minnesota, Minneapolis, MN, (2)Computer Science and Engineering, University of Minnesota, Minneapolis, MN, (3)Manufacturing Sciences and Technical Services, Genentech, Inc., Vacaville, CA

Multivariate data analysis has recently emerged as a critical tool to gain fundamental understandings of process characteristics, and thus assisting the search for possible intervention means to improve process performance. The time dynamics of 150 process parameters acquired from 243 production runs from the Genentech's Vacaville manufacturing facility were investigated in this study. Two complementary multivariate approaches, kernel-based support vector regression (SVR) and partial least square (PLS) regression, were used to predict process outcome indicated by the final antibody concentration and the final lactate concentration. Both productivity and cellular metabolic state can be predicted accurately in the early stage of the inoculum train, suggesting that the history of the culture exerts critical impact on the final process outcome. Furthermore, specific lactate production rate, viable cell density, and viability were identified as key parameters during the inoculum train contributing to the final process performance. In the production stage, most critical parameters were also related to lactate metabolism and cell growth, suggesting the important role of these two aspects of process characteristics. Interestingly, lactate consumption appeared to be a prominent factor in determining final process outcome, and was further investigated. This study represents an important step towards implementing PAT and QbD principles to enhance process understanding and facilitate dynamic control for improved process robustness.

See more of this Session: Applications of Multivariate Data Analysis In Pharmaceutical Process Development: Traditional and Advanced Multivariate Methods
See more of this Group/Topical: Topical I: Comprehensive Quality by Design in Pharmaceutical Development and Manufacture

(607a) CMC Reviewer Expectations for ANDAs Using QbD Approaches

Robert L. Iser, Division of Chemistry, FDA/CDER/Office of Generic Drugs, Rockville, MD

There has been an increase in science and risk-based approaches under the Quality by Design paradigm in new drug submissions to the FDA. There are particular challenges for generic drug manufacturers with respect to using QbD approaches, especially with more complex dosage forms, such as modified release (MR) products. In an effort to increase transparency with the generic industry and the pharmaceutical industry at large, this presentation will provide insight into CMC reviewer expectations when evaluating pharmaceutical development studies, the applicant's process and product understanding, use of risk assessment tools, and proposed control strategies.

See more of this Session: QbD for Modified Release Dosage Forms
See more of this Group/Topical: Topical I: Comprehensive Quality by Design in Pharmaceutical Development and Manufacture

MECHANISTIC MODELING AND COMPUTER SOFTWARE FOR QUALITY BY DESIGN IN FORMULATION OF MODIFIED RELEASE DOSAGE FORMS

Xiao Yu Wu[1] and Yousheng Zhou[2]
[1]Leslie Dan Faculty of Pharmacy, University of Toronto, [2]AP-CAD Inc., Toronto, ON, Canada

Introduction

Modified release dosage forms (MRDFs) can be designed and manufactured in various shapes and structures using different materials. Depending on the design, formulation and properties of drug and excipients (e.g. polymers), drug release from MRDFs can undergo varying mechanisms. In addition, the external conditions such as mixing degree, pH and volume of the releasing medium, may play an important role in the release kinetics of the drug. Therefore, thorough understanding of drug release mechanism of an MRDF and establishing a mechanistic model with consideration of important factors are essential for quality by design (QbD) in the development of MRDFs with target performance.

This presentation first describes the role of mathematic modeling and computer simulation in QbD, reviews the difference between mechanistic modeling and statistical modeling, and then discusses how mechanistic modeling and computer software can be employed for QbD in formulation of MDRFs using actual MDRFs as examples. Various multiparticulate systems are analyzed including matrix beads/pellets and minitablets with or without membrane coating, drug granules coated with polymer membrane, sugar pellets coated with drug layer and polymer coating. Prediction of *in vivo* performance from *in vitro* release profile and design of a MRDF from targeted *in vivo* plasma concentration profile will also be demonstrated.

The Role of Mechanistic Modeling and Computer Simulation in QbD for MRDFs

Although the concept of QbD in pharmaceutical product development and manufacturing has been proposed since 2004 by the FDA, mathematical tools for achieving QbD goals in MRDF design have not yet been fully developed. As illustrated in Figure 1, to implement QbD for development of pharmaceutical products, such as MRDFs, several steps are essential. Firstly we need to gain scientific understanding of how formulation and processing factors affect product quality and performance by theoretical analysis and experimental testing. Secondly we need a mathematic model with predictive power that is able to describe the relationship between formulation/process variables and product performance characteristics. Because of comlexicity of such relationship, sophisticated computer programming is required to make quantitative prediction. With the assistance of computer simulation, we are then able to design the formulation and process to assure product quality and desired attributes. Finally the performance of the products has to be confirmed by clinical studies. If the clinical results are satisfactory, the product development will be progressed to the next stage. Otherwise, the parameters in the model for the formulation and/or manufacturing will be adjusted. Modification of the model may also be required to take into consideration of new knowledge gained from the clinical studies. Therefore, mathematical modeling and computer simulation are useful tools for implementation of QbD of MRDFs.

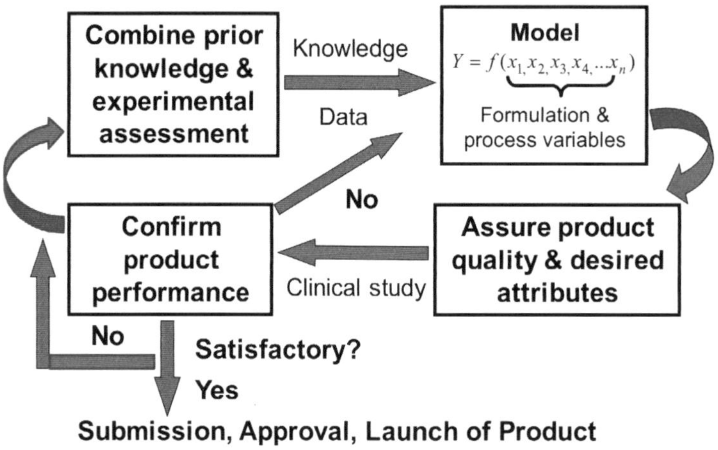

Figure 1. Schematic illustration of essential steps for implementation of QbD

Mechanistic Models versus Statistical Models

A variety of mathematical models and modeling methods can be employed to describe the relationship between formulation/process variables and product quality. Generally these models can be classified into two categories: (1) mechanistic models and (2) statistical models. The two types of models require different methodologies and input data sets and both have pros and cons as illustrated in Figure 2. For complex systems involving multiple variables simultaneously with unclear mechanistic relationships, statistical models exhibit advantages over mechanistic models. However, these models are post-experience models established by empirical correlation or statistical analysis and require a large amount of experimental data. To design MRDFs scientifically, mechanistic models are necessary as they can correlate excipient properties, drug properties, formulation composition, dosage form structure with the quality and performance of MRDFs based on underlying mechanisms and science.

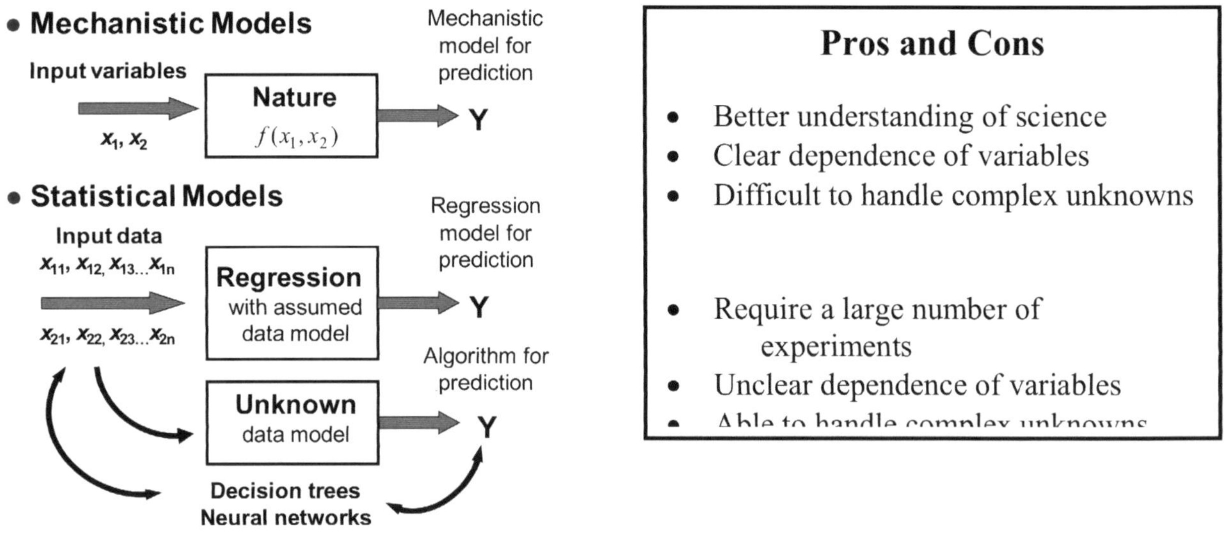

Figure 2. Comparison of mechanistic models and statistical models

Mechanistic Model-based Computer-Aided QbD of MRDFs

To build mechanistic models for MRDFs, we must first understand thoroughly the drug of interest (i.e., its physicochemical properties, pharmacology, pharmacokinetics, etc.), the drug delivery system and drug release mechanisms. Depending on the formulation and structure of MRDFs and drug properties, drug release kinetics may be controlled by drug diffusion, dissolution, polymer swelling/erosion, ion exchange, osmotic pressure, or a combination of several of these mechanisms. After having identified a rate-limiting release mechanism by theoretical analysis and preliminary experiments, we can derive a mechanistic model for the given drug and the delivery system. The model parameters, e.g. diffusion coefficient and partition coefficient, can be obtained by fitting experimental data of drug release with the model or determined by independent experiments or theoretical calculations (1-6). The experiment-verified parameters and model can then be applied to make predictions, to answer what-if questions, and ultimately to design MRDFs with desired quality attributes. In this presentation, we demonstrate the application of mechanistic model-based computer-aided QbD in typical polymer-coated multiparticulate dosage forms.

Multi-particulate systems such as ensembles of coated matrix beads/pellets or drug granules are extensively used for both systemic and local drug delivery. The multi-unit dosage forms are advantageous over single-unit ones in certain applications due to their flexibility in dosing and administration and low risk of dose dumping. In some formulations drug granules made of drug crystals are coated with film-forming polymers, and in some formulations a drug layer is sprayed onto sugar cores and then coated with a barrier polymer.

Application of Mechanistic Model-Based Computer-Aided QbD to Eudragit® Polymer-Coated Drug Granules

For a given drug, two formulation variables of coated drug granules, i.e., permeability and thickness of the coating, can be tailored to achieve target release profiles. EUDRAGIT® RL and RS polymers are acrylic polymers with different permeability due to a difference in the ionic monomer content which is higher in RL than in RS. They are widely used coating materials and can be mixed at any ratio to obtain desired permeability. In addition, one may apply different coating thickness of the polymers at a fixed RL/RS ratio to achieve desirable release profiles. However, it is quite challenging even for an experienced formulation scientist to decide a right thickness and RL/RS ratio to match a target curve for a given drug without multiple trial-and-error attempts.

Here we use theophylline as a model drug to demonstrate how mechanistic models and computer software can assist the design of EUDRAGIT® RL/RS coated drug granules. Based on the dissolution data in literature (7), the effective permeability of theophylline across coatings with various EUDRAGIT® RL/RS ratios was evaluated by using the optimization module in AP-CAD® software. The model-predicted release curves were then computed by using the obtained permeability, granule diameter of 0.55 mm, drug solubility $C_s \approx 0.01$ g/cm^3 and initial drug loading $C_0 \approx 1$ g/cm^3 (8). Figure 3 is a typical plot of model prediction compared with the test data. The correlation between theophylline permeability and the RL/RS ratio was used to predict release kinetics of granules with any given EUDRAGIT® RL/RS ratios. Figure 4 illustrates the effect of EUDRAGIT® RL/RS ratio on the release profiles for a fixed weight gain of 12%. Figure 5 portrays how weight gain of coating influences drug release profiles of granules at a given RL/RS ratio of 1:7. These examples demonstrated that formulation scientists can evaluate quickly different formulation options (i.e., RL/RS ratio) and physical parameters (e.g. coating thickness) to achieve target curves using this computational tool.

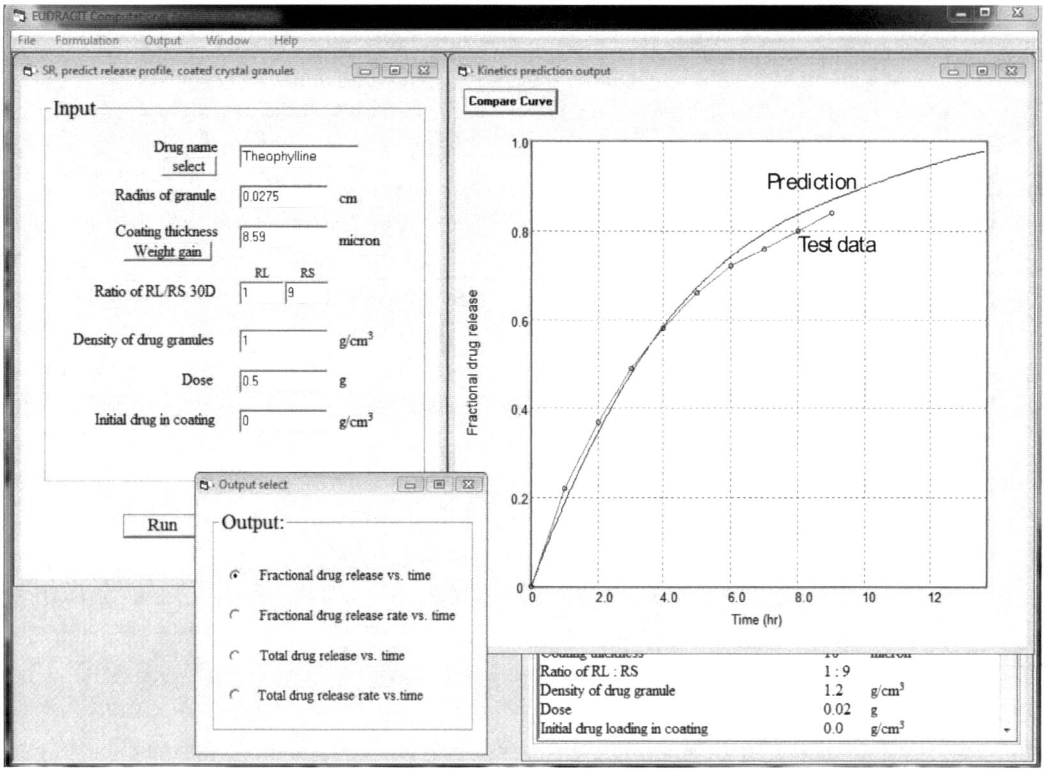

Figure 3. Comparison of model-predicted release profile with test data of theophylline granules

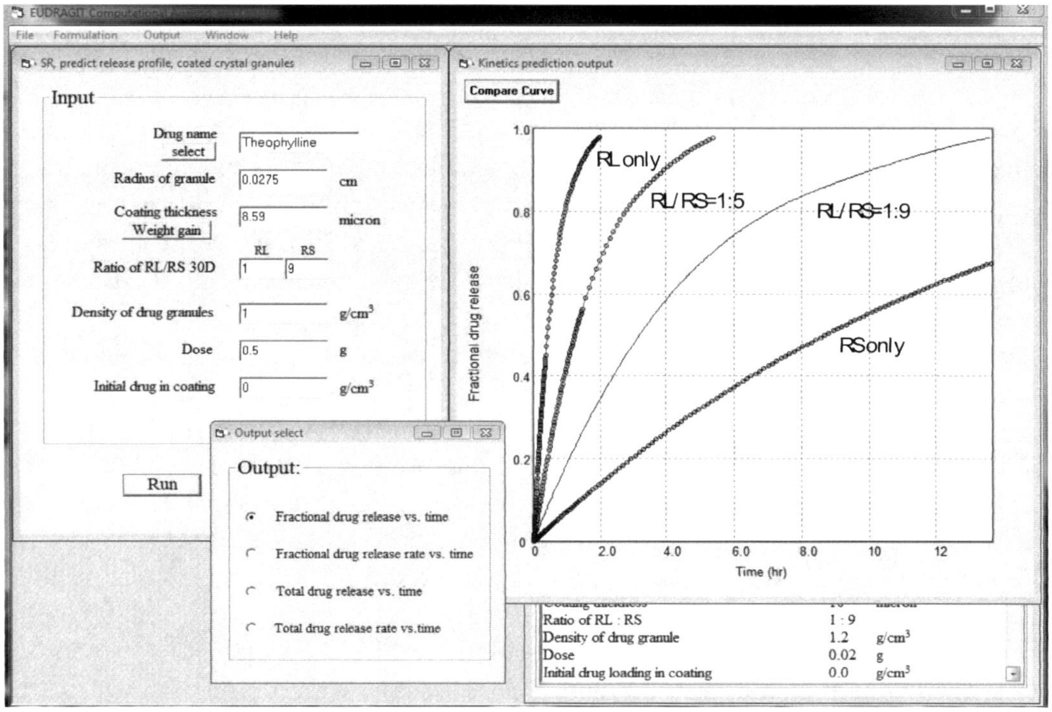

Figure 4. Effect of Eudragit® RL/RS ratio on theophylline release kinetics.

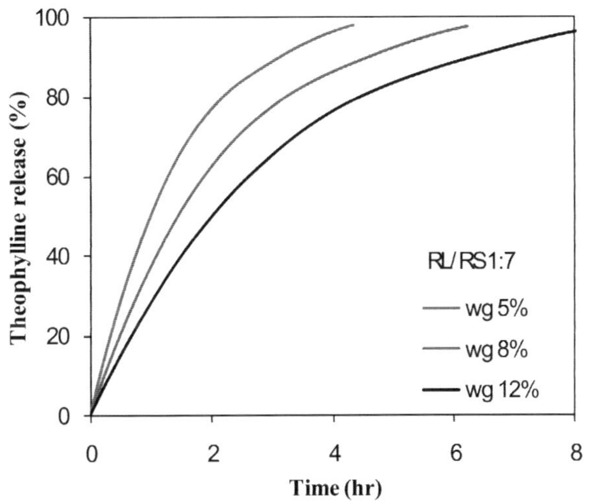

Figure 5. Effect of weight gain of coating on release profiles of theophylline granules coated with a Eudragit® RL/RS ratio of 1:7

Application of Mechanistic Model-Based Computer-Aided QbD to Surelease®-coated Drug Sugar Beads

Surelease® is an aqueous dispersion of ethylcellulose which is widely used as a barrier membrane in MRDFs such as coated beads and tablets. Since ethylcellulose is hydrophobic, a hydrophilic polymer (so called pore former), e.g. hydroxypropylmethylcellulose (HPMC) is often added in the dispersion to induce water-filled channels thus increasing drug release rate. Therefore, by adjusting the HPMC content in the coating, as well as the coating thickness, we are able to obtain different drug release profiles. A similar procedure of model building and verification presented above was used to study Surelease® coated metoclopremide HCl beads. Unlike the theophylline granules, this multiparticulate system has different structure and design. The drug was first sprayed onto sugar beads (nonpareil beads) and then coated with Surlease® E-7-7050 with or without pore former HPMC-E15 (9). A mechanistic mode was thus built with consideration of such structure and design.

Figure 6 portrays a typical plot of model-predicted release profile of metoclopremide HCl on sugar beads coated with Surelease® E-7-7050 and 5% pore former HPMC-E15 compared with test data (9) (the curve with circles). The model parameters were estimated from experimental data from Ref. 9 and the line was predicted by computer simulation using AP-CAD® software package for coated multiparticulate dosage forms. The verified model parameters, mathematical model and computational procedures were then used to analyze formulation and design factors. The influence of concentration of the pore former on the release kinetics is revealed in Figure 7. The impact of weight gain after coating (a measure of coating thickness) on drug release profiles at a given concentration of pore former (5%) is shown in Figure 8.

The above results indicate that, using the computer simulation, we can predict drug release profiles for given pore former concentration and coating thickness. On the other hand, we can identify an optimal pore former concentration and/or coating thickness for a desired release profile. This computational procedure can be combined with convolution method and drug pharmacokinetic data to predict plasma concentration profile of the drug; or with deconvolution method to design a coating formulation and thickness for a target plasma concentration profile (10).

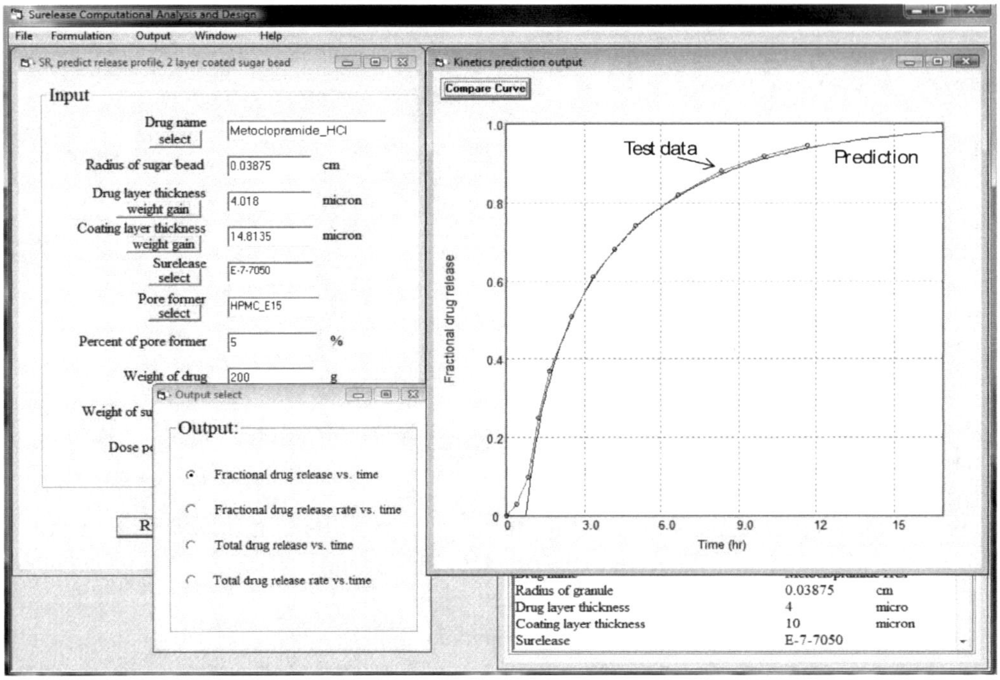

Figure 6. Comparison of model-predicted release profile with test data of metoclopremide HCl pellets

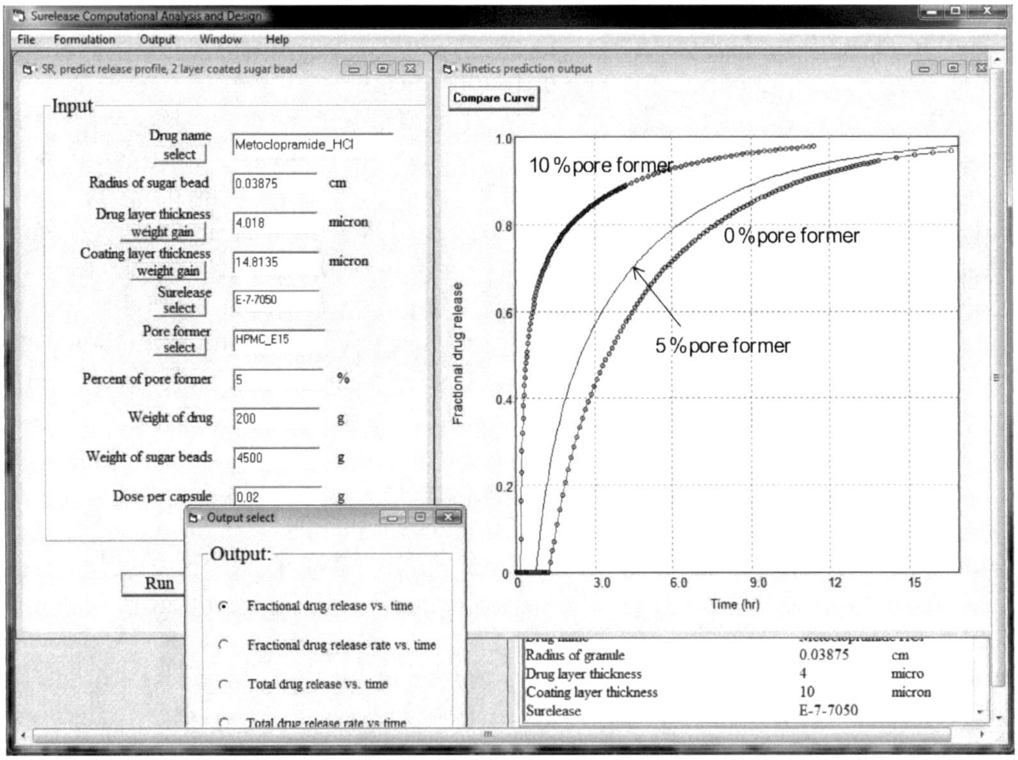

Figure 7. Effect of concentration of pore former HPMC-E15 on release profile of metoclopremide HCl pellets coated with Surlease® E-7-7050

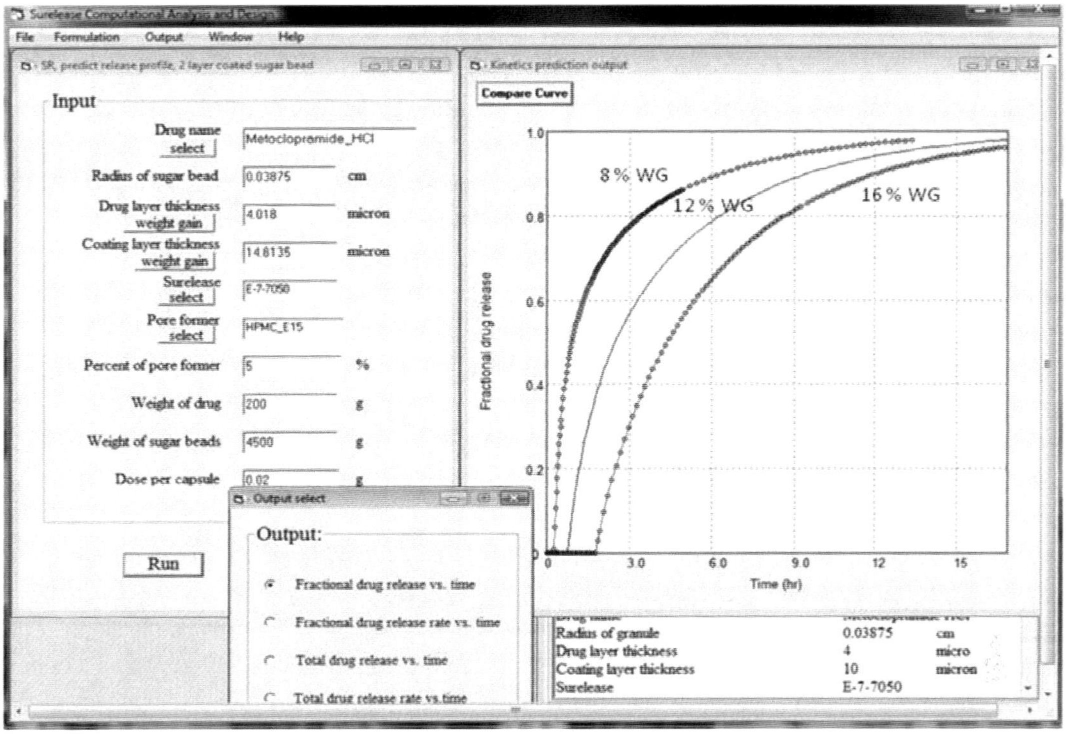

Figure 8 Effect of coating weight gain (a measure of coating thickness) on release profile of metoclopremide HCl pellets coated with Surelease® E-7-7050 and 5% pore former HPMC-E15

Conclusions

Mechanistic models and computer software are useful tools for QbD in formulations of MRDFs, as evident by the examples of EUDRAGIT® RL/RS 30D coated theophylline granules and Surelease® coated metoclopremide HCl sugar beads. This work has demonstrated that desired release profiles of various chemical entities can be achieved by altering physical parameters such as coating thickness (weight gain), and formulation parameters, e.g. EUDRAGIT® RL/RS ratio, or Surelease® and HPMC concentrations. The *in sillico* experiments can be accomplished in hours instead of days or weeks thus enabling scientists to explore greater formulation options and to define design space quickly.

Reference

1. Zhou, Y., Chu, J. S., Li, J. X., and Wu, X. Y. (2010), Theoretical analysis of release kinetics of coated tablets containing constant and non-constant drug reservoirs. *International Journal of Pharmaceutics, 385*(1-2), 98-103.
2. Zhou, Y., Chu, J. S., Zhou, T., and Wu, X. Y. (2005), Modeling of dispersed-drug release from two-dimensional matrix tablets. *Biomaterials, 26*(8), 945-952.
3. Zhou, Y., Chu, J. S., and Wu, X. Y. (2004), Theoretical analysis of drug release into a finite medium from sphere ensembles with various size and concentration distributions. *European Journal of Pharmaceutical Sciences, 22*(4), 251-259.
4. Wu, X. Y., Zhou, Y. (1999), Studies of diffusional release of a dispersed solute from polymeric matrixes by finite element method. *Journal of Pharmaceutical Sciences, 88*(10), 1050-1057.
5. Wu, X. Y., Zhou, Y. (1998), Finite element analysis of diffusional drug release from complex matrix systems. II. factors influencing release kinetics. *Journal of Controlled Release, 51*(1), 57-71.
6. Zhou, Y., Wu, X. Y. (1997), Finite element analysis of diffusional drug release from complex matrix systems. I. complex geometries and composite structures. *Journal of Controlled Release, 49*(2-3), 277-288.
7. Sinchaipanid, N., Chitropas, P., and Mitrevej, A. (2004), Influences of Layering on Theophylline Pellet Characteristics, *Pharmaceutical Development and Technology*, 9(2), 163–170.
8. EUDRAGIT Application Guidelines, (2008) 10th edition.
9. Sadeghi, F., Ford, J. L., Rubinstein, M. H. and Rajabi-siahboomi, A. R., (2000), Comparative study of drug release from pellets coated with HPMC or Surelease, *Drug Development and Industrial Pharmacy*, 26(6) 651-660.
10. Dutta, S., Qiu, Y., Samara, E., Gao, G., Granneman, G. (2005), Once-a-day extended-release dosage form of divalproex sodium III: Development and validation of a level A in vitro–in vivo correlation (IVIVC), *Journal of Pharmaceutical Sciences*, *94*(9), 1949-56.

(607c) Mechanistic Investigation of Drug Release From Asymmetric Membrane Tablets

Mary T. am Ende, Pharmaceutical Development, Pfizer Worldwide Research and Development, Groton, CT

An asymmetric membrane (AM) tablet was developed for a soluble model compound to study the in vitro drug release mechanisms in challenge conditions, including osmotic gradients, concentration gradients, and under potential coating failure modes. Porous, semipermeable membrane integrity may be compromised by a high fat meal or by the presence of a defect in the coating that could cause a safety concern about dose-dumping. Drug release from the AM system studied was determined to be robust to varying and extreme challenge conditions. The conditions investigated included varying pH, agitation rate, media osmotic pressure, media saturated with drug to eliminate the concentration gradient, simulated high fat meal, and intentionally placed film coating defects. Osmotic and diffusional shut off experiments suggest that the mechanism governing drug release is a combination of osmotic and diffusional at approximately 90-95% and 10-5%, respectively. In addition, the coating failure mode studies revealed this formulation and design is not significantly affected by a high fat meal or by an intentionally placed defect in the film coating, and more specifically, did not result in a burst of drug release.

See more of this Session: QbD for Modified Release Dosage Forms
See more of this Group/Topical: Topical I: Comprehensive Quality by Design in Pharmaceutical Development and Manufacture

(607d) QbD for Generic MR Drug Products

Yue (Helen) Teng, Office of Generic Drugs, FDA, Rockville, MD

The Office of Generic Drugs ensures the therapeutic equivalence and quality of generic drug products. This is based principally upon a regulatory paradigm that requires pharmaceutically equivalent drug products containing the same active ingredient, dosage form, and strength (among other things), and being bioequivalent to the brand name product. Also under this paradigm, the batch to batch quality of generic drug products is ensured based upon batch release testing according to specifications. While this approach has been used with great success toward bringing the availability of high quality of generic drug products, the vast majority of pharmaceuticals approved using this framework have been solution and immediate release oral products, which are inherently simple in design. Drug products however, have increased in design complexity, to encompass modified (oral) release products, among others complex dosage forms. As these pharmaceuticals have increased in their design complexity, approaches adopted to ensure the equivalence and quality of generic drug product pharmaceuticals, must likewise evolve.

FDA's Pharmaceutical cGMPs for the 21st Century: A Risk-Based Approach and Process Analytical Technology Initiatives are meant to enhance and modernize the regulation of pharmaceutical manufacturing and product quality. As a result of these initiatives and the trend of increasingly complex pharmaceutical products, OGD/GPhA has developed a mock example of pharmaceutical development report for a modified release product. This report illustrates how ANDA applicants can move toward implementation of quality by design (QbD). Through establishing quality target product profile (QTPP), identifying critical quality attributes (CQAs), performing risk assessment, conducting design of experiment, and setting control strategies, quality of generic modified release products is ensured.

See more of this Session: QbD for Modified Release Dosage Forms
See more of this Group/Topical: Topical I: Comprehensive Quality by Design in Pharmaceutical Development and Manufacture

FLUID BED COATING PROCESS FOR MODIFIED RELEASE APPLICATIONS

Orapin Rubino, Ph.D., Glatt Air Techniques, Inc., Ramsey, New Jersey

Biography

Orapin Rubino is Director, Formulation and Product Development, Pharmaceutical Services Division at Glatt Air Techniques, Inc. Her experience focus on development of solid dosage forms (sustained and modified release formulations, granulation, direct pelletization, pellet coating, drying, encapsulation and tableting processes). She received her B.S. degree in Pharmacy from Thailand and Ph.D. degree in Pharmaceutics from UNC Chapel Hill. She then spent two years as a postdoctoral research associate in Protein Chemistry (physical biochemistry) at Duke University, NC. Prior to joining Glatt in 1998, she worked as a Senior Process Scientist and Development Scientist at Dade Behring (formerly DuPont Medical Products) in Delaware for about 5 years. Her responsibilities at DuPont included formulation development, scale up and technical support for granulation and tableting of clinical diagnostic reagents.

Abstract

Wurster process (bottom spray fluid bed coating process) is a process of choice for coating of multiparticulates to produce modified or controlled release dosage forms. The unique features of the Wurster process include air distribution plates (up bed and down bed plates) and partition that organize fluidization of particles through the "coating zone". The nozzle is mounted at the bottom of a product container. The short distance between coating materials and particles during the coating process and the concurrent spray (same direction of particle flow and spray droplets) minimize spray drying. This principle contributes to high coating uniformity and coating efficiency of the Wurster process. This process can be used to produce pellets by drug layering process, follow by polymer coating to obtain modified or controlled drug release.

The fluidization pattern of particles is controlled by air distribution plates, partition height, and process air volume. Coating process parameters that mainly control the process include inlet air temperature, atomization air pressure/volume, and spray rate. Formulation aspect such as composition and solid content of the coating preparation can affect the drug release from the pellets.

Fluid Bed Coating

Diagram: Bottom Spray Coater (Wurster Process)

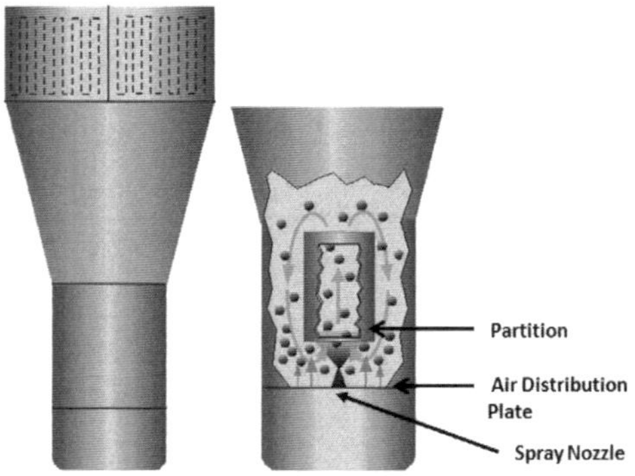

The principle and advantages of the Wurster process will be presented. Equipment, coating process parameters and scale up considerations will be discussed.

(607f) Implementation of QbD In the Development of a CR Dosage Form

Nancy E. Sever, Global Formulation Sciences, Abbott Laboratories, Abbott Park, IL, **Michael Brian Mackaplow**, Global Formulation Sciences- Solids, Abbott, Abbott Park, IL and Didier LeFebvre, Abbott, Abbott Park, IL

Use of QbD in characterization and development of a modified release dosage form will be discussed. Special emphasis will be placed on the early stages of development and the utilization of material sparing, strategic experimental design to ascertain the critical parameters affecting the performance of the dosage unit. Lab to Pilot to Commercial scale transfer will be discussed in terms of a QbD based approach. In addition to formulation development, opportunities for QbD in analytical method development will also be presented.

Additional information will be provided upon legal approval - End of May.

See more of this Session: QbD for Modified Release Dosage Forms
See more of this Group/Topical: Topical I: Comprehensive Quality by Design in Pharmaceutical Development and Manufacture

(607g) Control Strategy for Manufacturing of Modified Release Capsules: A Regulatory Perspective

Zhigang Sun, Office of Generic Drugs, OPS, CDER, FDA, Rockville, MD

Compared to simple dosage forms, there will be more challenges for modified release dosage forms to assure high drug product quality as well as manufacturing consistency due to its complexity of formulation design and process development. Quality by design (QbD) provides an effective tool to systematically evaluate, understand, and refine the formulation and manufacturing process to ensure predefined product quality. In this presentation, the modified release capsule manufacturing process will be used as an example to illustrate how to implement the QbD principles to enhance process understanding and to improve process controls in order to provide assurance of reproducible commercial manufacturing for consistent product quality. The purpose of this talk is to provide a regulatory perspective with regard to the QbD based control strategy for manufacturing of modified release drug products.

See more of this Session: QbD for Modified Release Dosage Forms
See more of this Group/Topical: Topical I: Comprehensive Quality by Design in Pharmaceutical Development and Manufacture

Managing Uncertainty in Design Space

Pat Whitcomb
Stat-Ease, Inc.
2021 E. Hennepin Ave
Suite 480
Minneapolis, MN 55413
Ph 612.746.2036
fax 612.746.2056
pat@statease.com

Mark Anderson
Stat-Ease, Inc.
2021 E. Hennepin Ave
Suite 480
Minneapolis, MN 55413

2011 AIChE Annual Meeting

AIChE 2011 1

Managing Uncertainty in Design Space

Objective: Define an operating window in process factor space where we have 95% confidence that 99% of the population meet (or exceed) specifications.

- Use empirical DOE to model the responses as functions of the process factors.
- Use a tolerance interval to "back off" (provide a buffer) from the specifications.
- Size the DOE for required half-width of tolerance interval.

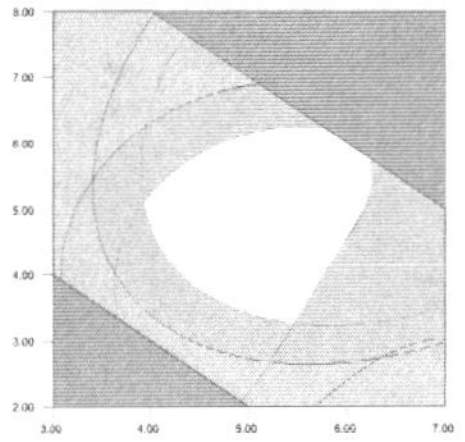

AIChE 2011 2

Illustrative Example
Tableting Process

This case study illustrates how response surface DOE and tolerance intervals can be used to set an operating window where specifications are consistently met.

- An optimal design is run on two process parameters (granulation time and lubrification time) in a tableting process.
- Three responses (dissolution, friability and hardness) are measured.
- The specifications are:
  - Dissolution ≥ 75%
  - Friability ≤ 0.5%
  - Hardness ≥ 10 kP

AIChE 2011 3

Initial Operating Window
Specifications as Bounds

y_1 = Dissolution %
(specification ≥ 75%)

y_2 = Friability %
(specification ≤ 0.5%)

y_3 = Hardness kP
(specification ≥ 10 kP)

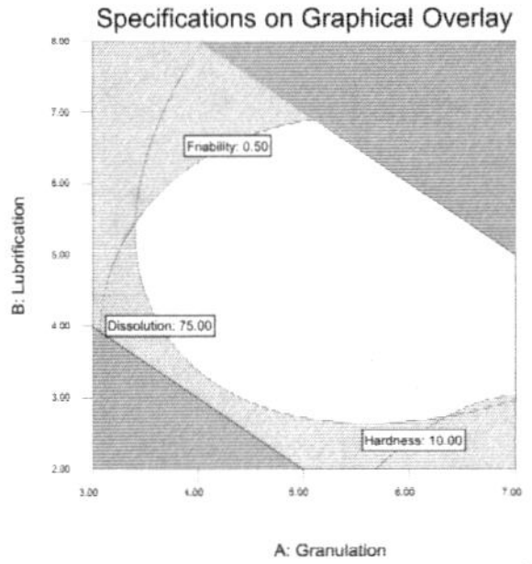

AIChE 2011 4

Uncertainty is a BIG Problem
Specifications as Bounds

If the tableting process is operated on a boundary, then 50% of tablets produced fall outside the specification.

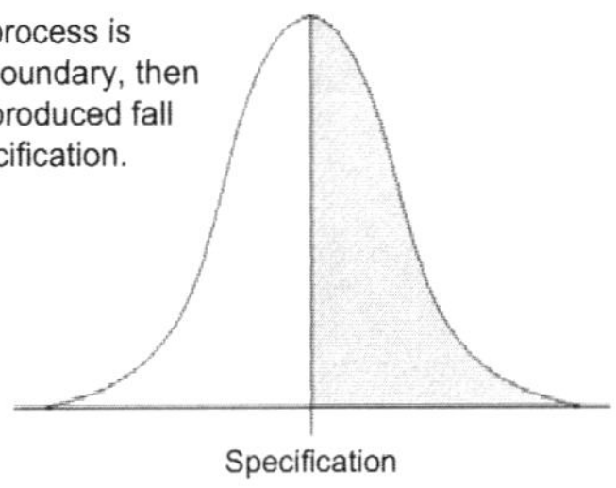

AIChE 2011 5

Managing Uncertainty in Design Space.

If we want more confidence that individual units are within specifications:

- Then we should back off using a tolerance interval rather than a confidence interval.
- Specify the confidence level; e.g. 95%.
- Specify the portion of the population to be within specifications; e.g. 99%.

Note:

- *A confidence interval (CI) pads our design space to give us confidence that the process mean is within boundaries.*
- *A tolerance interval (TI) give us confidence that a stated proportion of the population is within specifications.*

AIChE 2011 6

Tolerance Interval
Definition

Tolerance Interval: An interval that covers a fixed proportion of outcomes from the population with a pre-determined confidence level for estimating the population mean and standard deviation.

$$\bar{y} - k_2(s) \le P\% \text{ of population} \le \bar{y} + k_2(s)$$

$$TI = \bar{y} \pm k_2(s)$$

$$TI = \bar{y} \pm d$$

d is the half width of the tolerance interval

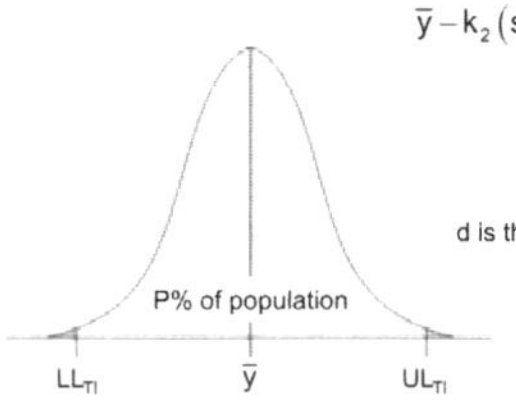

Interval Calculations
Single Sample & Normal Distribution

Confidence Interval: $\bar{y} \pm t\left(\frac{s}{\sqrt{n}}\right)$

Tolerance Interval: $\bar{y} \pm k_2 s$

Note:

- TI multipliers (k_2) are larger than those for a CI (t); especially for small samples.
- As the sample size increases:
  - CIs shrink towards zero.
  - TIs tend towards a fixed value: z(s).

df	$t_{\alpha=0.05}$	$k_{2\ \alpha=0.05,\ P=0.95}$
1	12.706	37.674
2	4.303	9.916
3	3.182	6.370
4	2.776	5.079
5	2.571	4.414
10	2.228	3.259
25	2.060	2.612
50	2.009	2.375
100	1.984	2.232
∞	1.960	1.960

Larger sample sizes are needed for TIs than CIs.

Tolerance Interval
Calculation for a DOE

TI calculation for a Normal Sample

$$TI = \bar{y} \pm k_2 s$$

TI calculation for a DOE[3]

$$TI = \hat{\bar{y}}_0 \pm s\left[t_{\alpha,df}\sqrt{x_0^T\left(X^T X\right)^{-1} x_0} + \Phi^{-1}(P)\sqrt{\frac{n-p}{\chi^2(n-p,1-\alpha)}}\right]$$

$$TI = \hat{\bar{y}}_0 \pm (s)(t_{\alpha,df})(\text{TI multiplier})$$

$$TI = \hat{\bar{y}}_0 \pm d$$

d is the half width of the tolerance interval

TI Interval Multipliers
Single Sample versus Two-Factor DOE

Comparison*:

single sample $k_{2(\alpha=0.05,\ P=99\%)}$

versus

$(t_{\alpha=0.05,df})(\text{TI multiplier}_{\alpha=0.05,\ P=99\%})$

Take home message:
Tolerance intervals that cover a design volume are wider than those at a single point in space.

* Basis is average TI multiplier for a two-factor unconstrained IV optimal design for a quadratic model with 6+ model points, 5 lack of fit points and 5 replicates.

df	$k_{2\ \alpha=0.05,\ P=0.99}$	(t)(TIM)
10	4.277	5.013
11	4.150	4.860
12	4.044	4.739
13	3.955	4.636
14	3.878	4.532
15	3.812	4.460
20	3.577	4.155
25	3.432	3.966
30	3.344	3.826
50	3.121	3.518
100	2.933	3.224

*TIM = TI multiplier

Interval Multipliers for "s"
TI DOE, TI sample and CI sample

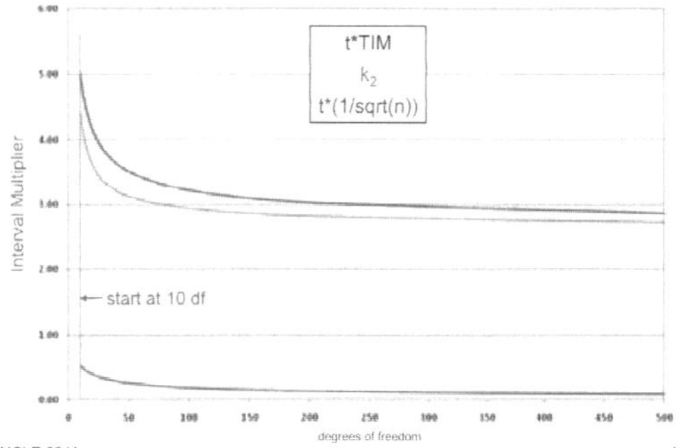

Back off from Specification
One-Sided 95% tolerance interval (TI)

One-sided TI calculation for DOE

$TI = \hat{\bar{y}}_0 + d$ or $TI = \hat{\bar{y}}_0 - d$

d is the offset for the tolerance interval

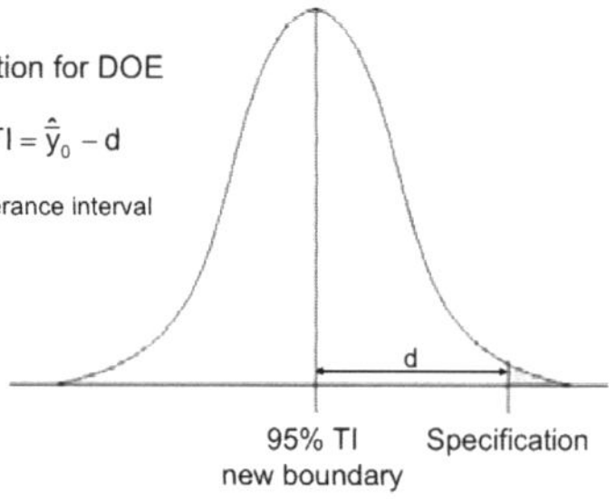

RSM DOE Process *(1 of 2)*
Tableting Process

1. Identify opportunity and define objective.
   Define an operating window in process factor space where we have 95% confidence that 99% of the population meets (or exceeds) specifications.
2. State objective in terms of measurable responses.
   a. Define the precision that is required for each response.
      Tolerance interval (α=0.05, P=99%) half-width (d):
      dissolution d = 9, friability d = 0.3, hardness d = 0.7
   b. Estimate experimental error (σ) for each response.
      dissolution s = 2, friability s = 0.03, hardness s = 0.14
   c. Use fraction of design space (FDS) to quantify precision.

RSM DOE Process *(2 of 2)*
Tableting Process

3. Select the input factors and ranges to study.
   (*Consider both your region of interest and region of operability.*)
   A = Granulation time: 3 – 7 minutes
   B = Lubrification time: 2 – 8 minutes
   7 ≤ A+B (total mixing time) ≤ 12
4. Choose a polynomial to estimate: *quadratic*
5. Select a design and:
   - Size design for precision needed *(next slide)*
   - Examine the design layout to ensure all the factor combinations are safe to run and are likely to result in meaningful information (no disasters).

DOE with Tolerance Intervals
Sizing for Precision Requirements

Define the precision required as the half-width of a Tolerance Interval that contains 99% of the population with 95% confidence:

Response	d	s	d/s
dissolution	9.0	2.00	4.5
friability	0.2	0.03	6.7
hardness	0.7	0.14	5

Size the DOE to the smallest (d/s) ratio.

Sizing for Precision Requirements
Tolerance Intervals *(page 1 of 3)*

16 runs (6 model, 5 lack of fit, 5 replicates)

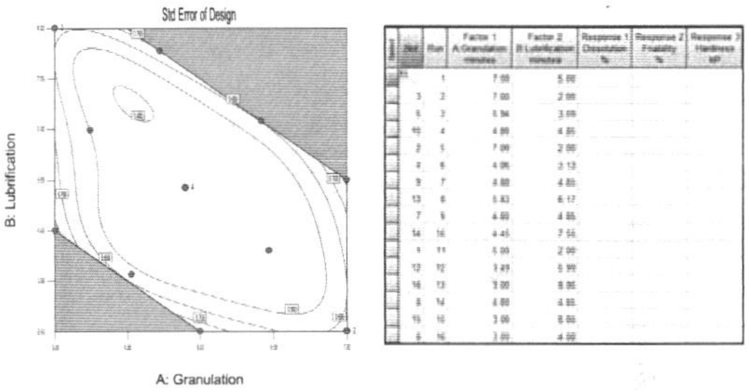

Sizing for Precision Requirements
Tolerance Intervals *(page 2 of 3)*

16 runs (6 model, 5 lack of fit, 5 replicates) FDS = 0%

d = 9
s = 2
a = 0.05
P = 0.99

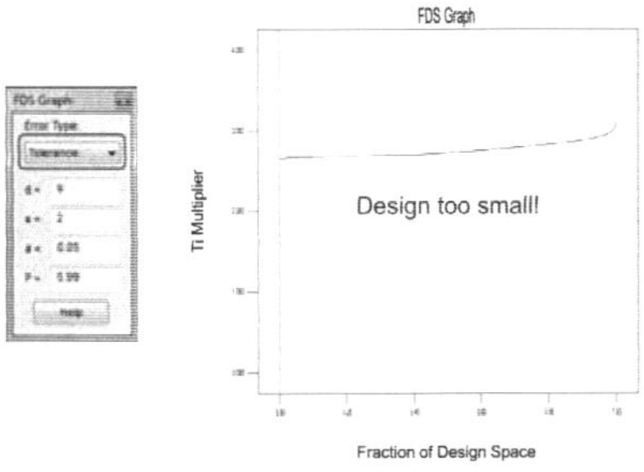

Sizing for Functional Design
Confidence Intervals

16 runs (6 model, 5 lack of fit, 5 replicates) FDS = 96%

d = 3
s = 2
a = 0.05
P = 0.99

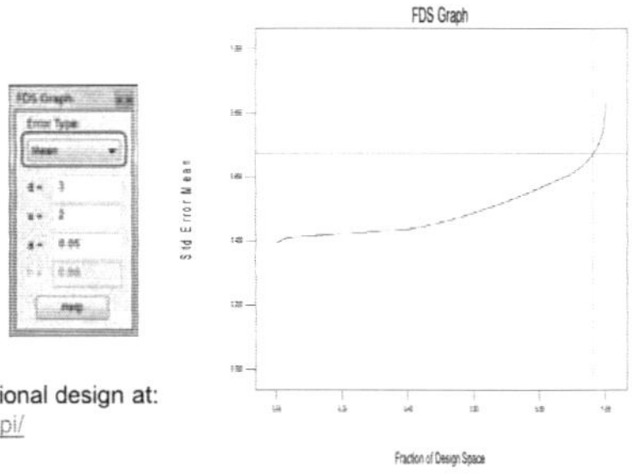

Sizing for functional design at:
http://asq.org/cpi/

Sizing for Precision Requirements
Tolerance Intervals *(page 3 of 3)*

23 runs (13 model, 5 lack of fit, 5 replicates) FDS = 96%

d = 9
s = 2
a = 0.05
P= 0.99

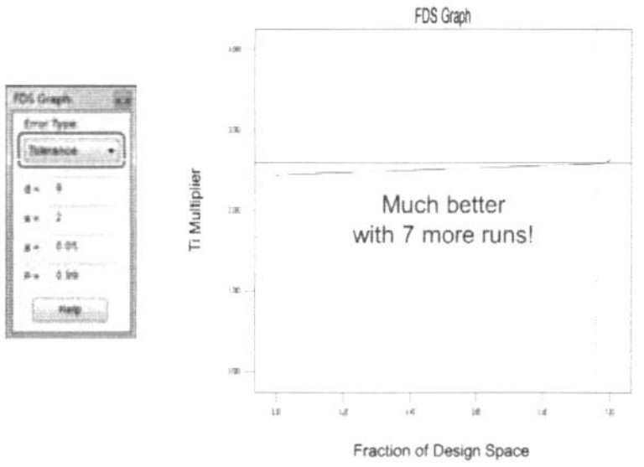

Tableting Process
Results *(page 1 of 2)*

Final design: 23 runs

- 13 model points
- 5 lack of fit points
- 5 replicates

Analyze responses, aim for:

- Dissolution ≥ 75%
- Friability ≤ 0.5%
- Hardness ≥ 10 kP

Std	Run	Factor 1 A:Granulation minutes	Factor 2 B:Lubrication minutes	Response 1 Dissolution %	Response 2 Friability %	Response 3 Hardness kP
	1	7.00	2.00	64.4	0.297	9.56
19	2	3.50	6.00	77.7	0.531	13.29
15	3	5.27	6.73	92.7	0.461	10.90
7	4	4.12	3.27	75.5	0.286	12.29
4	5	7.00	2.00	61.2	0.312	9.67
17	6	7.00	5.00	93.5	0.336	10.80
21	7	3.00	8.00	80.7	0.961	12.90
8	8	6.62	3.50	80.2	0.227	10.29
6	9	7.00	2.00	64.4	0.283	9.28
2	10	6.00	2.00	66.9	0.296	10.30
11	11	3.00	4.00	68.7	0.549	13.80
14	12	4.80	4.87	87.3	0.304	11.60
23	13	4.00	8.00	74.4	0.772	11.80
9	14	5.40	3.50	82.3	0.191	10.96
1	15	5.00	2.00	69.3	0.299	10.58
16	16	7.00	6.00	91.8	0.285	10.48
18	17	7.00	6.00	90.4	0.362	10.58
10	18	3.00	4.00	65.8	0.492	14.18
22	19	4.60	8.00	72.9	0.862	11.80
20	20	3.00	8.00	86.7	0.961	12.90
12	21	3.00	4.00	68.4	0.487	13.90
3	22	5.00	2.00	47.5	0.364	10.80
13	23	4.80	4.87	86.9	0.304	11.20

Tableting Process
Results *(page 2 of 2)*

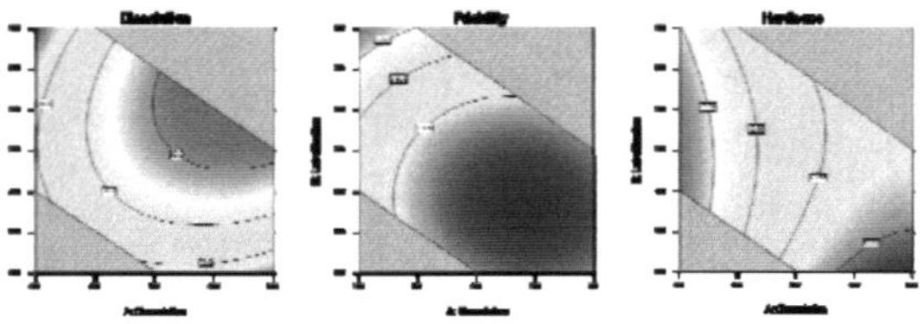

Final Operating Window
Tolerance Intervals as Bounds

y_1 = Dissolution %
(specification ≥ 75%)

y_2 = Friability %
(specification ≤ 0.5%)

y_3 = Hardness kP
(specification ≥ 10 kP)

Tolerance Intervals (α = 5%, P = 99%)

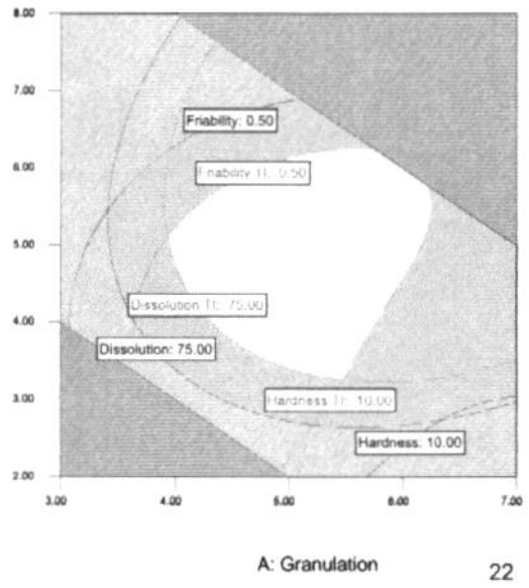

Quality by Design "QbD"
Design Space Determination

1. A region where quality product is produced.
2. Includes product formulation and/or process parameters.
3. Arrived at by iterative application of risk assessment and experimental design to knowledge space.
4. Includes evaluation of scale and equipment.

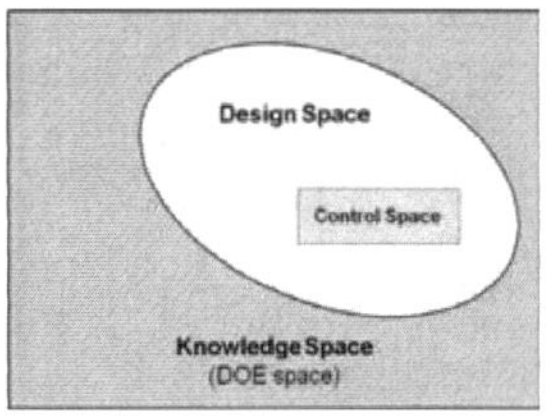

ICH guidelines Q8 (on Pharmaceutical Development), Q9 (on Quality Risk Management), and Q10 (on Pharmaceutical Quality System) provide some assistance for manufacturers to implement Quality by Design into their own operations.

Quality by Design "QbD"
Design Space Determination

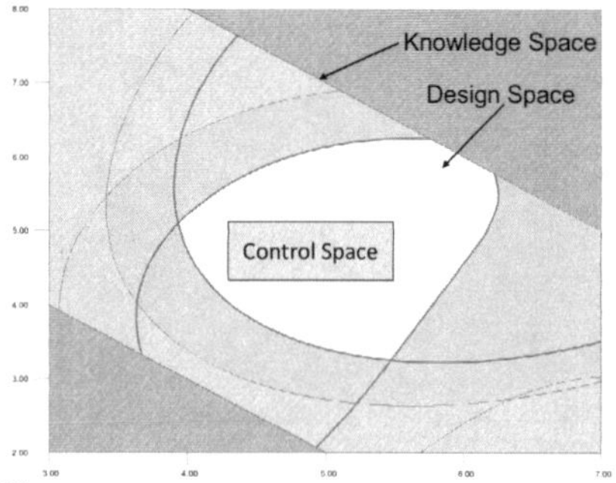

DOE with Tolerance Intervals
How many Runs

Quadratic model (TI for d = 9 and s = 2) FDS ≥ 90%

Factors	p + 5 + 5*	Extra runs	Increase
2	16	7	44%
4	25	10	40%
6	38	13	34%
8	55	15	27%
10	76	16	21%

* p coefficients + 5 lack of fit + 5 replicates

AIChE 2011 25

Managing Uncertainty in Design Space
Summary

1. Size your DOE appropriately for its propose:
   - Functional design – size via confidence intervals.
     A confidence interval pads our design space to give us confidence that the process mean is within boundaries.
   - Verify specifications – size via tolerance intervals.
     A tolerance interval give us confidence that a stated portion of the population is within specifications.
2. TI multipliers are larger than those for a CI.
   - As the sample size increases:
     - CIs shrink towards zero.
     - TIs tend towards a fixed value.
3. Tolerance intervals require a large design (sample size).

AIChE 2011 26

References

1. Alyaa R. Zahran, Christine M. Anderson-Cook and Raymond H. Myers, "Fraction of Design Space to Assess Prediction", *Journal of Quality Technology*, Vol. 35, No. 4, October 2003.
2. Heidi B. Goldfarb, Christine M. Anderson-Cook, Connie M. Borror and Douglas C. Montgomery, "Fraction of Design Space plots for Assessing Mixture and Mixture-Process Designs", *Journal of Quality Technology*, Vol. 36, No. 2, October 2004.
3. Steven DE Gryze, Ivan Langhans and Martina Vandebroek, "Using the Intervals for Prediction: A Tutorial on Tolerance Intervals for Ordinary Least-Squares Regression", *Chemometrics and Intelligent Laboratory Systems*, 87 (2007) 147 – 154.
4. Gerald J. Hahn and William Q. Meeker, *Statistical Intervals; A Guide for Practitioners*, (1991) John Wiley and Sons, Inc,
5. Gary Oehlert and Patrick Whitcomb (2001), Sizing Fixed Effects for Computing Power in Experimental Designs, *Quality and Reliability Engineering International*, July 27, 2001.

AIChE 2011 27

(655b) An Integrated Design Space Involving All Unit Operations in Tablet Manufacturing

Theodora Kourti, GSK, Hamilton, ON, Canada

A design space can be determined per unit operation, per cluster of unit operations or for the entire process.

By linking unit operations, and therefore recognizing a continuum in drug production that spans from the Drug Substance Properties to the Drug Product, it is possible to create a more versatile and robust Design Space. The final product that delivers the active pharmaceutical ingredient to a patient, is indeed the result of a multidimensional combination of raw material attributes and process parameters that span several unit operations including those of the drug substance production (like reaction and crystallization), the ones from drug product production (like granulation and compression) but also packaging. Each one of them can impact one or more final quality characteristics, as well as stability.

Therefore, if the Design Space is addressed in a holistic form, then the quality of the final product, including its stability, should be expressed as the result of the values and the interactions of these parameters. Also by treating the design space in a holistic way it would provide the manufacturer with the most cost efficient operation and provide for high yield and low operating costs, because problems at later unit operations would most likely be anticipated and corrected in earlier operations. In other words the control strategy will be part of the product design, such that it can be implemented in the most cost effective way. In this model, the tablet manufacturing process is mapped from start to end, and both design space and control strategy can be more dynamic as an observation at the extreme ends of the design space can be offset elsewhere in the process to provide conforming material.

This works gives an example where all unit operations are integrated in the Design space. Variables from all unit operations are incuded for continued process verification. The work includes different strengths of tablets

See more of this Session: Integrated QbD Approach for Design Space Development Including Real Time Release
See more of this Group/Topical: Topical I: Comprehensive Quality by Design in Pharmaceutical Development and Manufacture

(655c) Design Space Definition Using a Variational Bayes' Approximation

Linas Mockus[1], **Jos Miguel Lanez**[2], Lee E. Kirsch[3] and Gintaras V. Reklaitis[1], (1)Purdue University, West Lafayette, IN, (2)School of Chemical Engineering, Purdue University, West Lafayette, IN, (3) Division of Pharmaceutics, The University of Iowa, Iowa City, IA

The design space may be interpreted as the constrained region of the key manufacturing parameters space which provides assurance of quality of a drug given that manufacturing occurs in this region. The metric used to represent this assurance should not be deterministic, rather it should be stated in probabilistic terms. In this work, we report on the use of Bayesian methods to develop a suitable risk metric based on statistical models of the manufacturing processes and product properties. The Bayesian inference is carried out to determine the posterior distribution of the probability of the product meeting quality specifications. Here, we propose an alternative optimization based procedure, the variational Bayes' approximation, to obtain the posterior distribution. A sequential methodology is proposed to deal with non-linear models and the consideration of covariance. The results are compared to the widely used but computationally intensive Markov Chain Monte Carlo method. The proposed approached is illustrated with information drawn from a QbD study on Gabapentin.

See more of this Session: Integrated QbD Approach for Design Space Development Including Real Time Release
See more of this Group/Topical: Topical I: Comprehensive Quality by Design in Pharmaceutical Development and Manufacture

(655d) A Data-Driven Modeling Approach Using the Design of Dynamic Experiments Methodology In Calculating the Design Space of Batch Pharmaceutical Processes

Christos Georgakis, Systems Research Institute and Chemical and Biological Engineering, Tufts University, Medford, MA

The development of fundamental models for fed-batch pharmaceutical processes range from very challenging to almost impossible as not all the inner workings of the process are quantitatively known. Thus, the accurate calculation of the Design Space in which the process will be operated is a substantial challenge. One approach that has been recently proposed [1] is to use the Design of Experiments (DoE) methodology to develop data-driven models of the process. A major limitation of this approach is its inability to account for time varying operations that are often more optimal than those that are constant with time. Such time-varying operation include but are not limited to the varying the reaction temperature, the crystallizer cooling, the feeding rate of a sugar source to a fermentation. This dynamic character is even more important if the process will operate under feedback control. In such a case the operating conditions are changed by the controller to ensure the product quality at the end of the batch.

We present here an approach that utilizes the recently postulated Design of Dynamic Experiments (DoDE) methodology [2] that generalizes the classical DoE approach [3, 4]. This is achieved by systematically designing a set of experiments with respect to the time-varying decision variables. Using the collected data, a response surface model (RSM) is estimated that quantifies the impact that uncontrolled and controlled inputs (disturbances and manipulated variables, respectively) have on the quality of the product at the end of the batch. Process disturbances include the variability in feedstock properties as well as the variability in the operating conditions. The manipulated variables include those that can be changed with time such as the reactor temperature or the crystallizer cooling rate, or the substrate feed rate in fermentation, as well as others. A feedback controller aims to change the manipulated variables so that the variability in the final product is minimized.

The data-driven models derived from the DoDE data are used to quantify the design and control spaces of the process. Several simulated processes and an experimental asymmetric hydrogenation process of an API are examined to illustrate the general methodological concept that can be used in many batch pharmaceutical processes that avail the possibility of time varying conditions.

References:

1. A-Mab: a Case Study in Bioprocess Development CMC Biotech Working Group, Ver. 2.1

October 30, 2009, Internet Source: http://www.casss.org/associations/9165/files/A-Mab_Case_Study_Version_2-1.pdf
2. Georgakis, C. (2009); "A Model- Free Methodology for the Optimization of Batch Processes: Design of Dynamic Experiments" *Proceedings of the 8th IFAC International Symposium on Dynamics and Control of Process Systems (DYCOPS)* Istanbul, Turkey, July 2011
3. Box, G. E., & Draper, N. R. (2007); *Response Surfaces, Mixtures, and Ridge Analyses* (2nd Edition) Wiley- Interscience.
4. Montgomery, D. C. (2005); *Design and Analysis of Experiments* (6th Edition ed.) John Wiley & Sons, Inc.

See more of this Session: Integrated QbD Approach for Design Space Development Including Real Time Release
See more of this Group/Topical: Topical I: Comprehensive Quality by Design in Pharmaceutical Development and Manufacture

(655e) Dynamic Design Space As An Integrated Component of Quality by Design

Benoit Igne, **Carl A. Anderson** and James K. Drennen III, School of Pharmacy, Duquesne University, Pittsburgh, PA

The Food and Drug Administration requires from pharmaceutical companies the development of extensive knowledge databases regarding the control of the manufacturing process of a drug product. Through development and validation, drug manufacturers discover their process boundaries and identify a range of process parameters for each unit operation, the design space. However, limited work is done to study the effect of changing raw material on the robustness of the design space. Furthermore, once approved by the FDA, a process cannot be changed without filing post-approval changes that might require adding new experiments to the knowledge space. In the present study, the development of a design space for the creation of Excedrin (acetaminophen, caffeine) tablets through direct compression was investigated. A design of experiment including different excipent ratio of microcrystaline cellulose and lactose, two cros-carmelose sodium levels, and four compression forces was created using an industrial size press to define a knowledge space. Classical critical quality attributes (CQAs) (disintegration time, dissolution, radial tensile strength, friability) were measured and a design space identified. In order to test the robustness of the design space, raw material properties of acetaminophen (particle size) and lactose (ratio anhydrous to monohydrate) were modified. Resulting tablets were analyzed for relevant CQAs (disintegration time and radial tensile strength) to investigate how common variability in the raw material impacts the design space.

See more of this Session: Integrated QbD Approach for Design Space Development Including Real Time Release
See more of this Group/Topical: Topical I: Comprehensive Quality by Design in Pharmaceutical Development and Manufacture

(655g) Dissolution Testing of Prednisone and Salicylic Acid Calibrator Tablets At Different Tablet Locations

Anandhavalavan Arulmozhi and **Piero M. Armenante**, Otto H. York Department of Chemical, Biological and Pharmaceutical Engineering, New Jersey Institute of Technology, Newark, NJ

Dissolution testing is routinely carried out in the pharmaceutical industry to determine the rate of dissolution of solid dosage forms. This test is one of the several tests that pharmaceutical companies typically conduct on oral dosage formulations (e.g., tablets) to determine compliance. The USP Dissolution Testing Apparatus 2 is the most common of the apparatuses listed in the USP. However, it has been shown previously that the dissolution profile of a tablet undergoing dissolution in the USP Dissolution Apparatus 2 can be affected by the tablet location in the apparatus.

In this work, the dissolution rates of both non-disintegrating tablets (salicylic acid) and disintegrating tablets (Prednisone) were experimentally determined for many different tablet locations, both centered on the vessel bottom and off-center. The location of the tablet was experimentally varied in very small increments in order to determine the exact location where a transition in the dissolution profile occurred. It was found that in a small region (2-4 mm in radius) centered around the vessel centerline just below the impeller the dissolution profiles were similar to those observed with a centered tablet. However, outside this region the dissolution profiles were found to be significantly different, as indicated by the values of the Similarity Factor f_1 and the Difference Factor f_2. These finding are consistent with previous hydrodynamic investigations that showed the existence of a poorly mixed zone below the USP Apparatus 2 impeller. The results of this work can guide the practitioner on when to accept dissolution testing results based on tablet location.

See more of this Session: Integrated QbD Approach for Design Space Development Including Real Time Release
See more of this Group/Topical: Topical I: Comprehensive Quality by Design in Pharmaceutical Development and Manufacture

(696a) Use of Single-Use Technologies In Downstream Purification of Mabs - A Case Study

Venkatesh Natarajan, EMD Millipore, Billerica, MA

The employment of single-use technologies in the downstream processing of biologics in general and monoclonal antibodies in particular is gaining significant traction in the industry. Their use eliminates cleaning validation requirements and can potentially offer significant savings in capital costs and startup time, especially when designing new facilities. Though the advantages of single use technology have been widely talked about, there is relatively little application data, especially with respect to the entire downstream purification train. In this presentation, data will be presented on an entire mAb downstream purification train at 50 – 100 g scale, executed largely using single-use technologies including bags, tubing and filtration assemblies, devices and systems. Data will also be presented on the process development work performed to size the various unit operations.

See more of this Session: Innovations In Biopharmaceutical Processing: Factory of the Future
See more of this Group/Topical: Topical I: Comprehensive Quality by Design in Pharmaceutical Development and Manufacture

(696b) Alternative Approaches to Cell Culture Harvests

Gregory Frank, Amgen, Thousand Oaks, CA

Alteranatives to traditional centrifuge harvests for cell culture harvests are discussed.

See more of this Session: Innovations In Biopharmaceutical Processing: Factory of the Future
See more of this Group/Topical: Topical I: Comprehensive Quality by Design in Pharmaceutical Development and Manufacture

(696c) Scale and Bioreactor Design Translation for Rocking Bag Bioreactors

John Bowers, BioProcess Development, Merck, Union, NJ

Scale and Bioreactor Design Translation for RockingBag Bioreactors.

The translations of scale and bioreactor designs for disposable rockingbag bioreactors are currently performed intuitively or with trial an errorapproaches. Conditions are typically basedon manufacturer recommendations or historical usage. A momentum based fluiddynamic analysis of rocking bag bioreactors provides the foundation forestimating the mixing conditions, the gas transfer rates and the shearenvironment in the bioreactor. Few assumptions are needed, and fundamental,well established understandings of fluid flow provide quantitative descriptionsof various phenomena.

This work focused on WAVE

The results obtained from the approach was compared with literaturedata and observations of rocking bag bioreactors from 2-L to 500-L, includethose used in manufacturing. The analysispredicted well most important bioreactor aspects of bioreactor performance,including oxygen transfer limitations with scale, mixing issues with fillvolume, similarity of conditions to stirred tank bioreactors and micro-carrierattached cell damage. With thisunderstanding of the fluid mechanics, performance variations during scale-upand site changes can be anticipated and addressed, and proper operatingparameters can be set.

Development scientist andengineers who are using, or are evaluating rocking bag bioreactors. Any one who needs to understand the operatingprinciples of thistype of disposable bioreactor.

Education:

Ph. D. ChemicalEngineering, Princeton University; B. S. Chemical Engineering, Carnegie-Mellon University

Experience:

18 years ofbio-pharmaceutical equipment and process design, process scale-up and processmodeling experience at Merck / Schering-Plough in cell culture and downstreampurification.

See more of this Session: Innovations In Biopharmaceutical Processing: Factory of the Future
See more of this Group/Topical: Topical I: Comprehensive Quality by Design in Pharmaceutical

(696e) PAT Applications In Recombinant Protein Production Cell Culture Processes

Veronica Carvalhal, Lscc-PTD, Genentech, South San Francisco, CA

The regulatory agency's intent to make Process Analytical Technology (PAT) an industry-regulatory agreement makes this a very challenging and exciting time for both the pharmaceutical and biotechnology industries. In order to achieve the desired process understanding and control, the necessary effort should be used in developing tools and applying them either in the development or industrial stages. PAT is not a new strategy for the biotechnology industry or cell culture processes. Research and process development teams have long invested resources and expertise in process understanding. Nevertheless, PAT is a unique strategy where the final process adapts in real-time has not been applied frequently. This may be due to concerns with the challenges of implementing post-market process changes or costly and complex new technologies. In recombinant protein production processes, product quality is typically determined primarily by the selected cell clone and fermentation process conditions. Thus, upstream PAT process monitoring and control strategies may significantly benefit recombinant protein production processes. Fermentation pH or pO_2 controls are examples of PAT applications that have been applied and improved in the last couple of decades. More difficult to apply PAT strategies involve new sensor technologies or complex feedback approaches. Examples of nutrient controlled feeding and innovative air sparging strategies will be described as PAT applications that permit better process understanding and product quality control.

See more of this Session: Innovations In Biopharmaceutical Processing: Factory of the Future
See more of this Group/Topical: Topical I: Comprehensive Quality by Design in Pharmaceutical Development and Manufacture

(696f) At-Line NIR Spectroscopy As a Simple and Effective PAT Monitoring Technique In Mab Cultivations During Process Development and Manufacturing

Christian Hakemeyer[1], Silke Werz[1], Jose Cardoso Menezes[2], Francisca Folque[3], Gledson Jose[3] and **Ulrike Strauss**[1], (1)Pharmaceutical Biotech Production & Development, Roche Diagnostics GmbH, Penzberg, Germany, (2)Chemical & Biological Engineering Department, I.S.T., Lisbon, Portugal, (3) Technology Services & Products, 4Tune Engineering Ltd, Lisbon, Portugal

The experience of using at-line NIR transmittance spectroscopy on supernatant samples from a Mab cultivation process spanning several scales, clones and development years, from 2.5 l to 10,000 l is described. The collected spectra after preprocessing were used to (1) do process state estimation and (2) to obtain several culture parameters. Multivariate process trajectories were computed from NIR spectra acquired at-line and were used to: enhance process understanding across different scales up to industrial scale, assess batch-to-batch variability and scale-up issues, and examine the relative importance (criticality) of different sources of variability. The proposed approach demonstrates that NIR monitoring carried out at-line on samples already taken for reference methods analysis, (1) is very effective in replacing in-process monitoring of some critical process parameters by reference methods, (2) enables batch trajectories as defined by multiple culture parameters to be computed, (3) can be used to do guided sampling (i.e., near harvest or to investigate a deviation from the nominal trajectory by established bioanalytical methods) and (4) has potential use in implementing process corrections or even control strategies. The proposed approach is applicable throughout process development as well as to existing large-scale validated CHO bioprocesses, doesn't require more complex and expensive set-ups (viz., in-situ, on-line probes and process equipment) and is capable of supporting continuous improvement efforts on legacy and new but non-QbD Mab manufacturing processes.

See more of this Session: Innovations In Biopharmaceutical Processing: Factory of the Future
See more of this Group/Topical: Topical I: Comprehensive Quality by Design in Pharmaceutical Development and Manufacture

Challenges, Opportunities, and Scientific Aspects of Implementing PAT in Biopharmaceuticals

Huiquan Wu and Mansoor Khan, Food and Drug Administration, Silver Spring, MD
(*Corresponding author: Huiquan.wu@fda.hhs.gov; Tel: 301-796-0022)

Abstract: As an essential milestone for both the cGMP for the 21st Century and the recent pharmaceutical quality regulatory initiatives (Q8/Q9/Q10), the 2004 FDA Guidance for Industry "PAT—A Framework for Innovative Pharmaceutical Development, Manufacturing, and Quality Assurance" has proposed a science-based and risk-based regulatory pathway for pharmaceutical industry. Over the past 7 years the pace of implementation of PAT in small molecule drugs and biopharmaceuticals has been quite different. In this presentation, challenges and opportunities, technologies available as well as some scientific aspects of PAT real time process monitoring and control in biotech and biopharmaceutical area will be discussed based on several available case studies. In authors' opinion, when the pharmaceutical community is ready to embrace the challenges with enthusiasm, the benefits from implementing PAT in the biotech and biopharmaceutical area can be realized in the coming years.

1. Introduction

The pharmaceutical community has been facing significant challenges, such as increasing complexity not only in the aspect of product design such as dosage form and drug delivery system designs, but also in the aspect of pharmaceutical process design, drug manufacturing science and technologies. This requires more sophisticated and better controlled manufacturing process than simple, immediate release dosage form manufacturing process. The traditional pharmaceutical manufacturing mode has been batch processing, which relies on some tests of in-process samples and full tests of final products. This traditional mode, while can provide quality products to the consumers, it has some inherent disadvantages. For example, in the event a product issue is identified via final product testing, typically it will be hard to find out the root cause. Even if the root cause can be identified, it will be too late as the batch is completed by the time the analysis results become available. As late Nobel Laureate Dr. Werner Heisenberg said: "a single point can only tell so much; it doesn't give direction if only a single value is available". In this regard, real-time process monitoring and control offers significant advantages, as it can provide a scientific basis for timely critical process decision making and ultimately ensure that product quality is built-in.

As an essential milestone for both the cGMP for the 21st Century and the recent pharmaceutical quality regulatory initiatives (Q8/Q8(R2)/Q9/Q10)[1-4], the 2004 FDA Process Analytical Technology (PAT) Guidance[5] for Industry "PAT—A Framework for Innovative Pharmaceutical Development, Manufacturing, and Quality Assurance" has proposed a science-based and risk-based regulatory pathway for pharmaceutical sector. Over the past 7 years, it is noticeable that the pace of implementation of PAT in small molecule drugs and biopharmaceuticals has been quite different. In this paper, challenges and opportunities, technologies available as well as some scientific aspects of PAT real time process monitoring and control in biotech and biopharmaceutical area will be discussed based on several available case studies.

2. Rationale Scientific Considerations during Developing a PAT-Based Monitoring and Controlling System for a Pharmaceutical Process

The FDA PAT Guidance defines PAT as "a system for designing, analyzing, and controlling manufacturing through timely measurements (i.e., during processing) of critical quality and performance attributes of raw and in-process materials and processes, with the goal of ensuring final product quality. It is important to note that the term analytical in PAT is viewed broadly to include chemical, physical, microbiological, mathematical, and risk analysis conducted in an integrated manner. The goal of PAT is to enhance understanding and control the manufacturing process, which is consistent with our current drug quality system: quality cannot be tested into products; it should be built-in or should **be** by design." When developing an integrated PAT approach for any pharmaceutical processes including biotech and biopharmaceutical processes, as a process engineer, it is important to keep the following questions in mind:

(1) What to measure?
(2) Where to measure?
(3) How to measure?
(4) How to analyze the data?
(5) How to develop and validate a model?
(6) How to control the process in a timely manner?
(7) Etc.

Although this list can be easily getting long, based on our limited knowledge, the first three questions are fundamental—they are more measurement process design-oriented. It is imperative for these types of questions answered during the initial planning stage of a PAT-oriented project. The remaining questions are more chemometrics and control-oriented, which requires significant expertise and resources to dig in. However, they could be addressed during the 2nd stage of the PAT-oriented project.

First, we need to decide what property/quality attribute or process parameter to be measured or analyzed. For a typical PAT project, often times the final decision to the first question is either a critical quality attribute (CQA), or a critical process parameter (CPP), or a combination of CQA and CPP. The answer to this first question will be also linked to what kind of analytical technology to be used. Then we need to think about where to measure the property/quality attribute/process parameter in the process. This will be linked again to the choice of analytical technologies and process analyzers, and what location the process analyzer will be put as well. The choice of measurement location should be both representative and informative. Thirdly we need to figure out how to measure the property of the process material and/or process parameter. Depending on the way the process and/or process equipment the process analyzer is interfaced to, the answer to this 3rd question might be at-line, in-line, on-line, or even near real time as defined in the FDA's PAT Guidance. With the first three questions answered, the answers to the questions remaining on the above list will be more technical-oriented. Therefore, answering these questions would require significant resources and expertise.

3. Challenges and Opportunities for implementing PAT in biotech and biopharmaceutical area

Over the last 7 years, PAT development and implementation in the small molecule pharmaceutical area has been making steady progress. The general philosophy of PAT has been incorporated into a broader pharmaceutical quality initiative "Pharmaceutical Quality for the 21st Century Initiative", or Quality-by-Design (QbD). The essential PAT elements, including PAT process monitoring and control, process understanding, have been become key QbD elements. According to ICH Q8 and Q8(R2), process control strategy is required for a successful QbD implementation. In a broad sense, the process understanding and process knowledge gained through PAT efforts, as well as control strategy established via PAT implementation, will collectively eliminate or mitigate the risk of delivering pharmaceutical products with poor or substandard quality to the patients and the public. Therefore, PAT is a scientific mechanism for pharmaceutical risk management. The general risk management framework is outlined in the ICH Q9 guideline.

Compared to their small molecule counterparts, biotechnology products and biopharmaceutical processes are recognized by their complexity due to facts including but not limited to, (1) variability existing in both product and process, such as compositional and structural heterogeneity, lot-to-lot variability; (2) process outcomes sensitive to foreign perturbations and process conditions, etc. Therefore, implementation of PAT in the biotech and biopharmaceuticals area would normally require more efforts and dedication. In addition, collaboration among various disciplines is expected to play a significant role.

Depending on the types of critical quality attributes (CQAs) and critical process parameters (CPPs) to be monitored and controlled, different technologies or process sensors could be used for implementation of PAT for a biotech or biopharmaceutical process. Table 1 below lists several technologies used for small molecule PAT/QbD applications based on reference[6] and authors' experience. Some of them are applicable to biotech and biopharmaceutical PAT/QbD applications, too. Taking the cell culture unit operation as one example, Table 2 blow is a list of technologies available for process monitoring and control, not meant to be an exhaustive list (adopted from reference[7]).

Table 1 Several Technologies used for small molecule PAT/QbD applications (some of them applicable to biotech and biopharmaceutical PAT/QbD applications, too)

Techniques	Quantitative Data	Qualitative Date	Raw material ID	Physical Parameters	On-Line Capability	Near-Line Capability
Mid-Infrared	Yes,with ATR and probes; usually only liquid samples	Yes, with ATR and probes; powder and liquid samples	Yes, not easily but similar to qualitative data	Yes, but not as clean as NIR	Yes, but more difficult than NIR/Raman; firbers expesnive and brittle	Yes, samples may be taken in near real time.

Near-Infrared	Yes, any physical form; transmission or refelection	Yes, large numbers of chemometric algorithms	Yes, all forms; ID and quality	Yes, hydrogen bonding makes it sensitive to all changes	Yes, either wireless units or by fiber probes from hardened units	Yes, ruggedized units are standard
Chemical Imaging	Yes, possible with chemometric algorithms	Yes.	Yes.	Yes, For example, hardness, domain size statistics, etc.	Not yet.. But possible. One system was in the pilot plant in 2003.	Not yet. Excellent capability for characterizing samples off-line (intact) quickly.
Far-Infrared or Terahertz	Possible with chemometric algorithms	Yes, sensitive to crystalline materials	Yes, but mostly for crystalline materials	Yes, intermolecular THz absorption makes it easy to detect interfaces or layer boundaries	Not yet. Recent development?	Not yet, But has excellent off-line characterization capability (both intact or destructive mode)
Raman	Yes, but semi-quantitative; without reference, not truly quantitative	Yes, insensitive to water	Yes, portable units are available	Yes, hydrogen bonding seen	Yes, mainly liquids and slurry systems.	Yes, both portable and traditional units
FBRM	Yes, but needs validation.	Yes.	Not applicable	Yes, chord length.	Yes, mainly liquids and slurry systems.	Yes, samples may be taken in near real time.
PVM	Yes.	Yes.	Not applicable	Yes, particle size and morphology.	Yes, mainly liquids and slurry systems.	Yes, samples may be taken in near real time.
In-situ AFM	Yes.	Yes.	Not applicable	Yes.	Yes, mainly solids or solids exposed to liquids	Yes.

Table 2 Possible process parameters to be monitored and/or controlled by various PAT sensors for a mammalian cell culture unit operation (adopted from reference[7])

Process parameter	**PAT tool**	**PAT application example**
Glucose concentration	Continuous FIA in conjunction with chemiluminescence assay	Control volume mixing pattern and residence time monitored to indicate metabolic condition to prevent cell starvation
Biomass concentration	Conductance and capacitance-based biomass probe	Specific growth rate and product formation monitored to determine onset of stationary phase and cell lysis
Biomass concentration	In-situ microscopes coupled with image processors and computer control	Specific growth rate used to control product formation in agitated cell broth cultivation
Oxygen uptake rate and specific oxygen uptake rate	Mass-balance technique for measuring kLa, dissolved O2 concentration, gas compositions	Control of physiological state of cells and cellular activity
Biomass, glucose, and acetate concentrations	NIR spectroscopy and electronic nose device	Control of fed batch cultivation and monitoring of the quality of the culture
Cell size, biomass concentration	Dielectric spectroscopy	Closed-loop control of medium feed rate in fluidized bed fermenters
Glucose and biomass concentrations	Multiwavelength fluorescence (MVF), software sensors and Multivariate Data Analysis	Control of specific growth rate and product quality
Gas and hydrocarbons	Gas sensor array as an electronic nose	Detection of early infections in mammalian cell cultures
Heat fluxes	Flow micro calorimeter, dielectric spectroscopy	Monitoring of specific growth rate and control of utilization of media components

It is appreciable that scientific researches in the area of biotech and biopharmaceutical PAT/QbD applications have been pretty active over the last few years. Some innovative pharmaceutical companies have also actively presented their biotech and biopharmaceutical PAT works at premier meetings such as IFPAC, AAPS Annual Meetings, AIChE Annual Meetings, etc. Most noticeably, the FDA CDER Office of Biotechnology Products (OBP) QbD pilot program[8] was launched in July 15, 2008. All of these efforts and programs together created a positive momentum for biotech and biopharmaceutical PAT/QbD to move forward. While the PAT/QbD initiative opens a new quality paradigm for biotech and biopharmaceutical development and manufacturing, numerous opportunities and challenges are co-existing. For example,

(1) How to use the PAT tools listed above and other new technologies to gain in-depth and better process understanding of the biotech and biopharmaceutical processes?
(2) How to use other modern analytical tools to gain better understanding of the product characteristics of the mAbs and other biopharmaceuticals?
(3) How to use advanced control tools to better control the biotech and biopharmaceutical process?
(4) How to use Design of Experiments (DOE) and other multivariate statistical tools to better understand and manage the variability inherent with biotech and biopharmaceutical products and processes?
(5) How to integrate the PAT/QbD methodologies into the existing subject knowledge system?
(6) How to establish the linkage among the process, product, and clinical safety and efficacy of drug products?

It calls for collaborative effort among the industry, academia, and regulatory parties. At the moment, our understanding of the link between each process parameter, each product quality attribute, and the clinical safety and efficacy of the drug is generally limited. It would be ideal if the linkage among the process, product, and patient can be established via an integrated manner in the PAT/QbD paradigm, although most likely it will be case-by-case.

4. Case Studies for Biotech and Biopharmaceutical PAT applications

For the two case studies illustrated below, some of the scientific aspects of implementing PAT/QbD in the biotech and biopharmaceutical area will be explored to certain detail.

4.1. NIR Monitoring of a Bioreactor Process (Adopted from Reference[9])

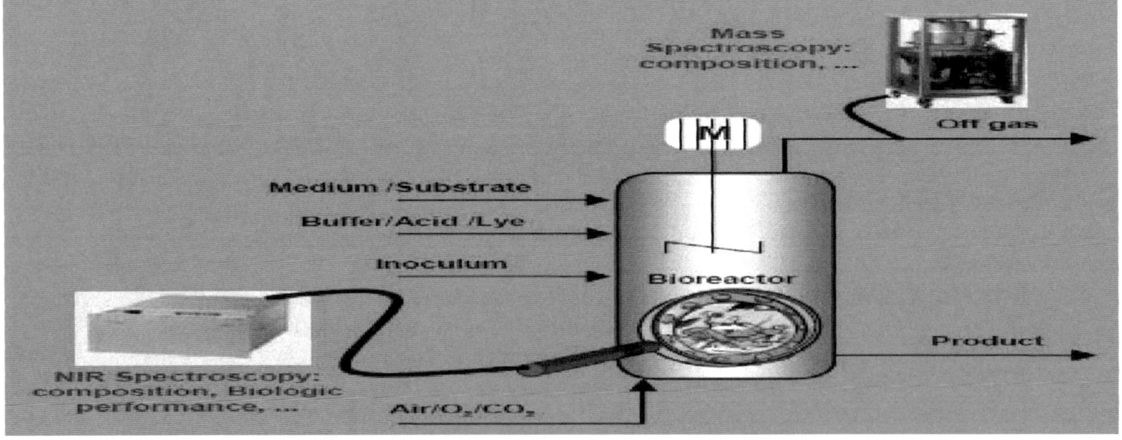

Figure 1. PAT real time monitoring of a bioreactor process

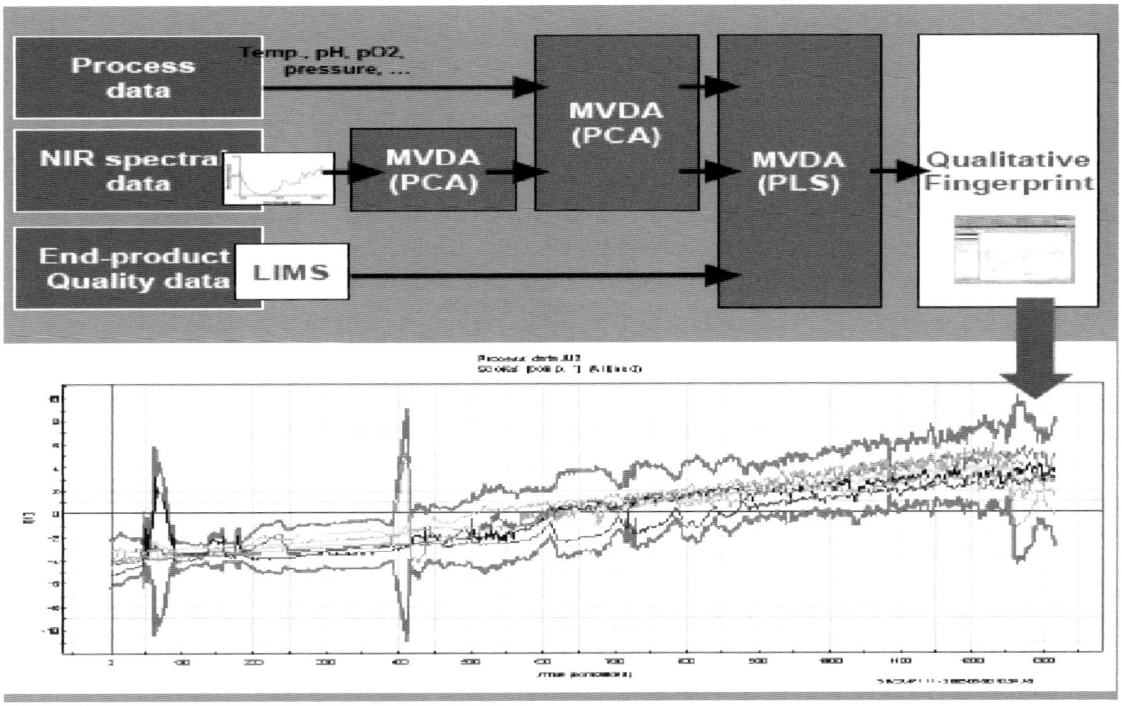

Figure 2. Data analysis and modeling strategy for a PAT-based bioreactor process

Although detailed data were not shown here, this publicly presented case study illustrated a PAT concept of using an in situ transmission probe and other process probes (pH, O_2, etc.) to online monitor a bioreactor process, and linking it to the end-product quality data via chemometrics techniques (such as Principal Component Analysis (PCA), Partial Least Square (PLS) regression). The chemometrics tools were used to identify qualitative fingerprint of the process, and to create a process trajectory for on-line process control. Similar case studies have been reported for some small molecule drug PAT applications by many researchers including authors[10-13].

4.2. Rapid Monitoring of Recombinant Antibody Production by Mammalian Cell Cultures Using Fourier Transform Infrared Spectroscopy and Chemometrics (Adopted from Reference[14])

In a recently published work[14], Fourier transform infrared (FT-IR) spectroscopy combined with multivariate statistical analyses was investigated as a physicochemical tool for monitoring secreted recombinant antibody production in cultures of Chinese hamster ovary (CHO) and murine myeloma nonsecreting 0 (NS0) cell lines. Medium samples were taken during culture of CHO and NS0 cell lines, and analyzed by FT-IR spectroscopy. It was found that FT-IR generates information on multiple aspects of the medium over the length of a culture, varying from utilization of the medium components to excretion/secretion of materials from cells into the medium.

The PCA shows that a clear trend in PC1 from left to right is observed for all cell lines. The loading plot PC1 shows that wave numbers corresponding to amide bonds (1700-1550 cm^{-1}) have a positive effect, where those corresponding to polysaccharides (1200-1000 cm^{-1}) have a negative influence. Therefore, PCA is separating the samples on the basis of both cell growth as indicated by the consumption of polysaccharides, and accumulation of antibody, as reflected by increased protein in the medium.

PLSR analysis was used to predict antibody titers, and these regression models were able to predict antibody titers accurately with low error when compared to ELISA data. PLSR was also able to predict glucose and lactate amounts in the medium samples accurately. The work demonstrates that FT-IR spectroscopy has great potential as a tool for monitoring cell cultures for recombinant protein production and offers a starting point for the on-line measurement of antibody production in industrial scale bioreactors.

In summary, FT-IR Spectroscopy as a means to monitor cells in culture has certain advantages and disadvantages. Advantages include: (1) Simultaneous assessment of mAb titer, glucose consumption, and lactose accumulation; (2) Minimal sample volume required (<20 μL); (3) Technically simple; it avoids need for chromatography. Disadvantages include: it can only scan dehydrated samples. For practical purposes, FTIR spectroscopy can be used for monitoring cell growth and determining optimal media conditions, and for selection of high-producing clones.

An internal study of using real time spectroscopy to monitor a bioreactor process has been undertaken within the FDA. Preliminary yet promising results have been obtained. However, the details of real time spectroscopy as a PAT tool for the process monitoring are to be worked out via further experiments and techniques such as chemometrics and other modeling methodologies. The study results will be released when ready.

5. Concluding Remarks

PAT is vital for implementing QbD in Pharmaceuticals, including biotech and biopharmaceutical processes and products. PAT focuses on process understanding, process control, process & product design. Chemometrics and process modeling are enabling tools to extract information and knowledge from process and product data, and to formulate and present the design space for regulatory fillings. In authors' opinion, when the pharmaceutical community is ready to embrace the challenges with enthusiasm, the benefits from implementing PAT in the biotech and biopharmaceutical area can be realized in the coming years.

Disclaimer: ***The views and opinions expressed in this paper are only those of the authors, and do not necessarily reflect the views or policies of the FDA***.

Acknowledgements:

The supports from colleagues and management at authors' division (Division of Product Quality Research) and offices(Office of Testing and Research, Office of Biotechnology Products, Office of Pharmaceutical Sciences) in CDER, FDA, as well as from various technology providers and authors of literatures cited are acknowledged.

References

1. FDA/ICH, 2006a. Guidance for Industry. Q8 Pharmaceutical Development. Available at: http://www.fda.gov/cder/guidance/6746fnl.pdf

2. FDA/ICH, 2009. Guidance for Industry, Q8(R2) Pharmaceutical Development. Draft available at:http://www.fda.gov/downloads/Drugs/GuidanceComplianceRegulatoryInformation/Guidances/ucm073507.pdf

3. FDA/ICH, 2006b. "Guidance for Industry. Q9 Quality Risk Management". Available at: http://www.fda.gov/cder/guidance/7153fnl.pdf

4. FDA/ICH, 2007. Q10. Pharmaceutical Quality System. Draft Consensus Guideline. Available at: http://www.fda.gov/cder/guidance/7891dft.pdf

5. FDA, 2004. "Guidance for Industry. PAT—A Framework for Innovative Pharmaceutical Development, Manufacturing, and Quality Assurance". Available at: http://www.fda.gov/cder/guidance/6419fnl.pdf

6. Ciurczak, E.W.. "A Spectroscopy Field Guide for PAT", Pharmaceutical Manufacturing, July/Aug. 2008, p.18-26

7. Rathore, A.S., Bhambure, R., Ghare, V. 2010. Process analytical technology (PAT) for biopharmaceutical products. Anal Bioanal Chem. DOI 10.1007/s00216-010-3781-x

8. Morris, W. and Hough, E., Parenteral Drug Association (PDA) Newsletter, "QbD Pilot Program for Biotech Seeks Similar Answers as One for Small Molecules", Nov-Dec, 2008.

9. Ingrid Maes. Application of PAT in bioreactor process development and manufacturing. the FIP 2008 PAT Workshop, Aug. 30-31, 2008. Basel, Switzerland.

10. Wu, H. and Khan, M.A., 2009. Quality-by-Design (QbD): An Integrated Approach for Evaluation of Powder Blending Process Kinetics and Determination of Powder Blending End-Point. J Pharm Sci. 98(8), 2784-2798.

11. Wu, H., Tawakkul, M., White, M., and Khan, M.A., 2009. Quality-by-Design (QbD): An Integrated Multivariate Approach for the Component Quantification in Powder Blends. Int J Pharm. 372(1-2), 39-48.

12. Wu, H. and Khan, M.A., 2010. Quality-by-Design (QbD): An Integrated Process Analytical Technology (PAT) Approach for Real-time Monitoring and Mapping the State of a Pharmaceutical Co-precipitation Process. Journal of Pharmaceutical Sciences 99(3): 1516-1534.

13. Wu, H. and Khan, M.A., 2011. Quality-by-Design (QbD): An Integrated Process Analytical Technology (PAT) Approach to Determine the Nucleation and Growth Mechanisms during a Dynamic Pharmaceutical Co-precipitation Process. Journal of Pharmaceutical Sciences 100(5): 1969-1986.

14. Sellick, C.A., et al., 2010. Rapid Monitoring of Recombinant Antibody Production by Mammalian Cell Cultures Using Fourier Transform Infrared Spectroscopy and Chemometrics. Biotech & Bioeng. 106(3):432-442.

(701a) Modeling a Reactive Crystallization with a Reduced Population Balance Equation for Dissolution and Growth to Control Critical Quality Attributes

Daniel Patience[1], Linli He[2], Dale Mowrey[2], Partha Mudipalli[3], Michael Christie[2] and Roger Bakale[2], (1)Chemical Process Research & Development, Cephalon, Malvern, SD, (2)Chemical Process R&D, Cephalon, Malvern, PA, (3)API Technologies, Cephalon, West Chester, PA

Modeling has always been a powerful weapon in a chemical engineer's arsenal. Use of modeling should be considered a key piece in the application of Quality by Design (QbD) in drug substance and drug product development. In this study, reaction, dissolution and growth kinetics are determined for a reactive crystallization of an intermediate-grade drug substance (DS) currently in development at Cephalon, Inc. Initial data showed that decreasing particle size of the starting material (in the DS formation step) increased the levels of entrapped starting material in the DS product. Smaller particles of starting material dissolve faster enhancing the rate of product formation. This leads to product coating the remaining starting material, thus increasing the level of entrapped starting material. The level of starting material in the DS is a critical quality attribute (CQA). Also, decreasing the particle size of the product led to lower levels of starting material entrapment. Breakage exposes the interior starting material to additional reagent. DS particle size is a potential CQA. The goal of this work is to model and hence predict entrapment levels and particle size as a function of dissolution, reaction, and crystallization kinetics. In the solution phase, a Mannich reaction occurs which is modeled as two second order reversible reactions in series. The model equations to describe this reactive crystallization consist of a solution phase mass balance for the reaction kinetics coupled with two population balance equations (PBEs). The PBEs describe the particle size densities for the nearly insoluble starting material and the nearly insoluble DS product during dissolution and precipitation. Due to the low solubility of the starting material and the DS product, we assume the solution phase concentration is always saturated with respect to these two materials. Therefore, a reduced fast dissolution and high growth model was developed to describe this process. The model in this study is a new model for reactive crystallizations because of the reduced nature and the ability to describe multiple crystallization phenomena and two second order reversible reactions in series with minimal parameters. Using the modeling software package Octave, the model parameters (reaction, nucleation and nuclei-disappearance rate constants) were estimated by fitting model predictions to experimental measurements including particle number (via Mettler's Lasentec FBRM probe), reactor temperature, and percent entrapped starting material in DS. This type of modeling approach is a macroscopic modeling approach for QbD and can be superior to a factorial or central-composite design approach because it can offer physical insights. Such insights include knowledge of the reaction and crystallization rate constants and the ability to extrapolate beyond previously studied experimental conditions.

(701b) Model-Based Characterisation of Organic and Aqueous Tablet Film Coating Processes: Parameter Estimation and Risk Management

Salvador Garca Muoz[1], Mary T. Am Ende[2], **Mark A. Pinto**[3] and Sean K. Bermingham[3], (1)Process Modeling & Engineering Technology, Pfizer Global Research and Development, Groton, CT, (2)Pharmaceutical Development, Pfizer Global Research and Development, Groton, CT, (3)Process Systems Enterprise, London W6 7HA, United Kingdom

Tablet coating is an important processing step in thepharmaceutical industry in which tablets are coated with an aqueous or organicfilm coating for both aesthetic and functional reasons.

Mathematical models of tablet film coating are important inpharmaceutical development as they aid in design of experiments, scale-up anddetermining optimal process conditions as formulations change. A universalsteady-state film coating model has previously been developed (am Ende,Berchielli, 2005) with the aim of providing process engineers with a means ofpredicting target operating conditions for optimization, scale-up androbustness studies. In that work, the model was validated with experimentalconditions and the predictions obtained were found to be in good agreement withdata not used for model validation.

In this study, the steady-state model was first validatedagainst steady state plant data using gSOLIDS (Process Systems Enterprise, UK).It was found that starting with one experiment, as experiments were added, the95% confidence interval decreased rapidly reducing to less than 10% with fouror more experiments.

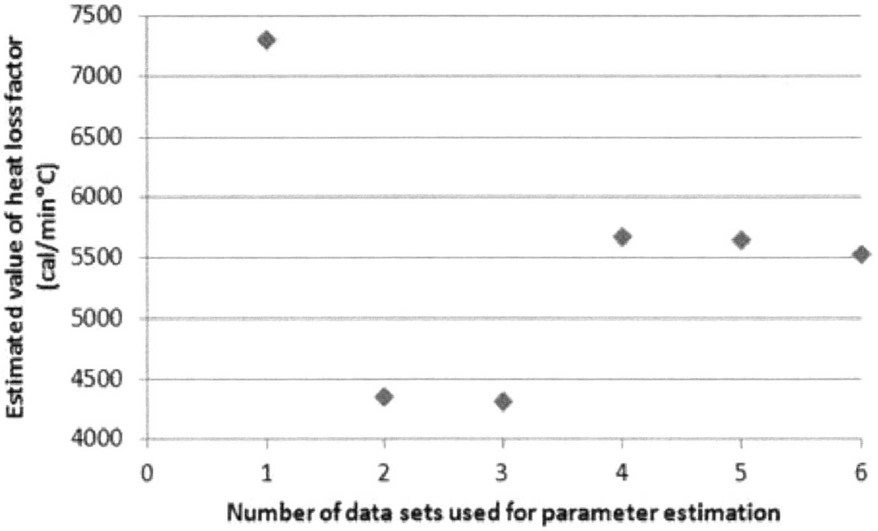

Figure 1: Change in estimated value of heat loss factor as experimental data is added

In order to quantify whether sufficient data was used formodel validation, the 95% T-value was compared to the reference T-value. (Ifthe 95% T-value is greater than the reference T-value, this indicates thatsufficient data was used to for model validation). The results obtainedindicated that at least three experiments are needed to accurately estimate theunknown model parameter. This was consistent with the results obtained for the95% confidence interval which was more than 50% when only two experiments wereused.

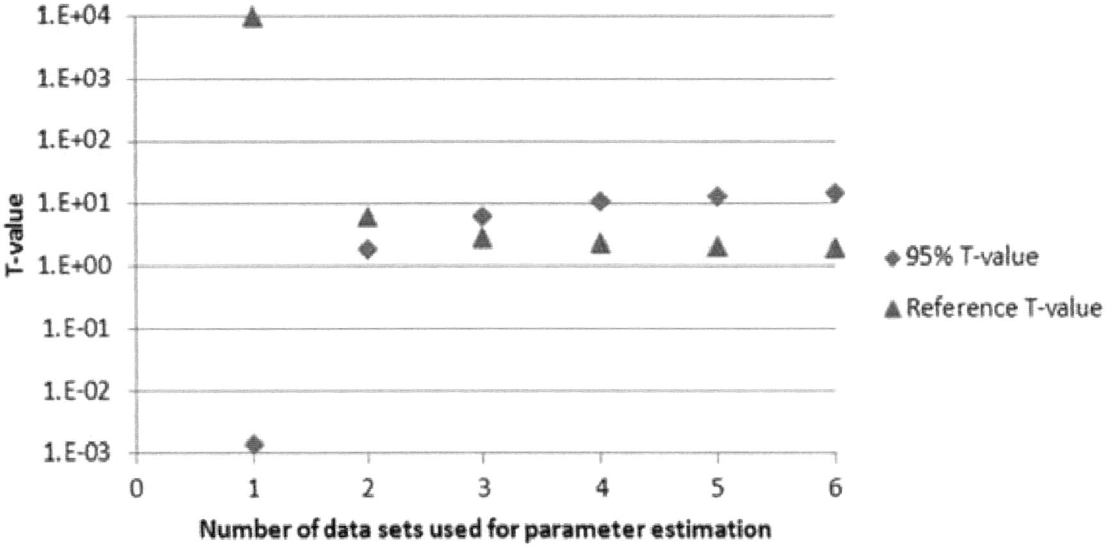

Figure 2: Changes in reference and 95% T-values as experimental data is added

As indicated by the 95% confidence intervals, the estimatedvalue of the model parameter is not a perfect estimate. Therefore in order todetermine a reasonable estimate of the model prediction a Monte Carlosimulation was carried out. In this simulation several instances of the modelwere run with each instance having the unknown model parameter sampled from anormal distribution centred at the parameter estimate and with a standarddeviation obtained from the 95% confidence interval. The results from thesenumerous simulations were then averaged to obtain an estimate of the modelprediction. In addition, the results from these simulations were used to obtaina 95% confidence interval on the model predictions. The results obtainedindicated that the model predictions were in good agreement with theexperimental data used for parameter estimation. Further, given the confidenceinterval of the estimated value of the model parameter, the standard deviationof the model prediction was much smaller than that of the data indicating thatthe predictions of the model vary very little with respect to the uncertaintyin the parameter estimate.

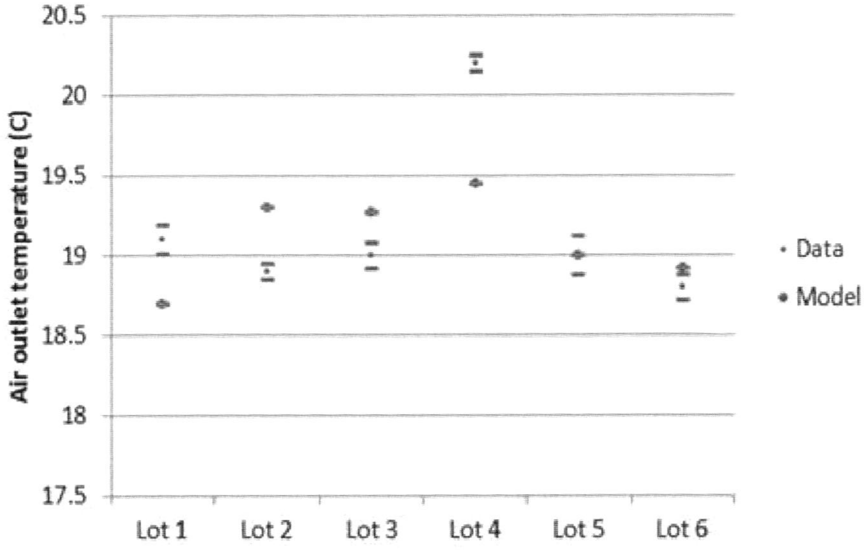

Figure 3: Comparison of model predictions with experimental data

A similar analysis was conducted using dynamic data. In thiscase however, although the data used for parameter estimation was deemedsufficient, the quality of the fit was quantified as being unsatisfactory evenif several experiments were used. This indicates insufficiencies in the modelwhich is expected as a steady state model was used to study dynamic plant data.

In summary, the results presented above indicate that withrelatively few steady state experiments, an accurate estimate can be obtainedof the unknown model parameter in the steady state film coating model. Themodel predictions were found to be in good agreement with the experimentaldata. The uncertainty in the model parameter was quantified and it was foundthat the model predictions were relatively unaffected given the uncertainty inthe parameter estimate.

The analysis presented above is generic and can be used toquantity uncertainty in any mathematical model. This information can beespecially useful as it can be used to quantify the risk associated withdecisions made using these mathematical models. Further, the rigour of theanalysis helps determine whether more experimentation is needed or whetherfurther model development is needed (or possibly a combination of both).

References

M. am Ende, A Berchielli (2005) A thermodynamic model fororganic and aqueous tablet film coating. *Pharmaceutical Development andTechnology*, 1:47-58

See more of this Session: Model Based QbD In Drug Substance or Drug Product Development
See more of this Group/Topical: Topical I: Comprehensive Quality by Design in Pharmaceutical Development and Manufacture

(701c) Design Space Verification for Tablet Dosage Form Using Modeling

Mary T. am Ende[1], Thomas P. Garcia[2], Kim Vukovinsky[3], Penelope Butterell[4], Alex Chueh[5], Bernd Schuemmelfeder[6], Vincent E. McCurdy[2] and Daniel Gierer[1], (1)Pharmaceutical Development, Pfizer Worldwide Research and Development, Groton, CT, (2)Global CMC, Pfizer Worldwide Research and Development, Groton, CT, (3)PharmaTherapeutics, Pfizer Worldwide Research and Development, Groton, CT, (4)Gqts, Pfizer Global Supply, Zaventem, Belgium, (5)Global Manufacturing Services, Pfizer Global Supply, Pearl River, NY, (6)Global Manufacturing Services, Pfizer Global Supply, Freiburg, Germany

Quality by Design principles are well established in the pharmaceutical industry. It is becoming common place to develop the design space for a drug product and include it in the regulatory filing. There are a broad range of methods to establish design space, some of which are focused on individual unit operations and others that take a holistic view of the entire product processing. A new challenge is developing approaches for verifying the design space for commercial production, especially for those defined using data obtained from modeling or small scale experiments. In this talk, a proposal for verifying the design space for a drug product with multiple CPPs, KPPs and CQAs will be discussed using principles of scale-dependencies, use of precedented models, and (when necessary) large scale confirmation batches.

See more of this Session: Model Based QbD In Drug Substance or Drug Product Development
See more of this Group/Topical: Topical I: Comprehensive Quality by Design in Pharmaceutical Development and Manufacture

(701d) New Model-Based Platform for Enhanced Content Uniformity Prediction of Oral Dosage Forms

Salvador Garca-Muoz, Process Modeling & Engineering Technology, Pfizer Worldwide Research & Development, Groton, CT and Weili Yu, Pharmaceutical Development, Pfizer Worldwide Research & Development, Groton, CT

New Model-based Platform for Enhanced Content Uniformity Prediction of Oral Dosage Forms

Being a critical quality attribute for capsules and tablets, content uniformity (CU) has received continued attention from practitioners and academics with multiples proposals to relate it with upstream properties such as the particle size distribution and true density of the active pharmaceutical ingredient (API).

In this work we propose a new model to predict the CU of a lot of tablets or capsules, based on a corrected algorithm that considers: the particle size distribution of the API and its variability, the true density of the API, the size dependant potency of any drug product intermediate and the weight variability of the lot of finished product. The model is based on a particle level mass balance and a Monte-Carlo simulation of the sampling process. The accuracy of this new model is demonstrated by showing predicted versus observed values of content uniformity (RSD) for multiples cases.

This model enables a much improved risk assessment exercise between API characteristics, production process and CU; and provides a platform to establish acceptable variability ranges and targets that are fundamental components of a Quality by Design development.

See more of this Session: Model Based QbD In Drug Substance or Drug Product Development
See more of this Group/Topical: Topical I: Comprehensive Quality by Design in Pharmaceutical Development and Manufacture

(701e) Computation of Reliabilities for Extent of Reactant Conversion Incorporating Batch Effects with Applications to ICH Q8 Design Space

Brian R. Crump[1], Rick Lewis[2], Zifang Guo[2] and John Peterson[3], (1)Process Engineering, GSK, Research Triangle Park, NC, (2)Statistical Sciences, GSK, Research Triangle Park, NC, (3)Quantitative Sciences, GSK, Upper Providence, PA

A fundamental reaction model was developed to explain rate observations under periods of both kinetic and mass transfer control. The reaction starting material was a potential genotoxin such that the residual starting material concentration in the product was strictly regulated. The primary means of controlling the starting material concentration in the product was via the extent of the reaction. Hence, it was important to be able to reliably forecast the reaction conversion as a function of process and equipment parameters. This presentation will demonstrate a means of gauging the reliability of the reaction model.

It has been shown (Peterson, 2008) that a predictive distribution for quality responses was important for process capability and ICH Q8 design space quantification. Experiments involving mechanistic models for API production often have relatively small measurement error variability. However, predicting a future quality response involved sources of variation other than just measurement error. For example, one needed to model the batch-to-batch variation, which can be substantially larger than the measurement error variation.

Within the Bayesian statistical framework, development of a predictive distribution for future quality responses is, in theory, straightforward. However, it can be computationally challenging to compute such distributions, and numerically intensive methods are often necessary. In this presentation, we use the BUGS software application to develop Bayesian predictive models with and without random variation due to batch effects. We show that failure to model the batch-to-batch variation can result in noticeable bias for process capability and ICH Q8 design space development.

See more of this Session: Model Based QbD In Drug Substance or Drug Product Development
See more of this Group/Topical: Topical I: Comprehensive Quality by Design in Pharmaceutical Development and Manufacture

(701f) Estimating Reaction Model Uncertainty with Bayesian Markov Chain Monte Carlo

Jacob Albrecht, Bristol-Myers Squibb Company, New Brunswick, NJ

The use of kinetic models to predict reaction performance is becoming an integral element of pharmaceutical process development. These models, built on fundamental mechanistic understanding and laboratory data, must be characterized by how well they can describe experimental observations. This task of quantifying uncertainty for reaction models is complicated by model non-linearity and experimental error. Because experimental error may not follow a conventional distribution, the validity of common statistical assumptions behind least-squares regression and normal error distributions should be reexamined when fitting models to data.

This paper reports a comparison of different techniques to compare parameter regression and confidence intervals for pharmaceutically relevant chemical reaction models. Simulated experiments were created by adding two types of noise (normally distributed and experimentally measured) collected using two analytical techniques (HPLC versus inline FTIR) to four types of kinetic models (linearized, exponential, coupled ODE and DAE). The best fit parameter values and confidence intervals were then calculated according to a multivariate *t*-distribution or two different Bayesian Markov-Chain Monte Carlo algorithms: one estimating error assuming a normal error distribution and a new, more adaptable non-parametric likelihood function. The uncertainty estimates were evaluated based on their ability to correctly regress the model parameter's probability distribution in scenarios common to pharmaceutical reaction engineering. The findings indicate that while multivariate *t*-distributions are sufficient for simpler non-linear models with normal error distributions, only BMCMC accurately estimates parameter ranges when errors are non-normal or when the model system is complex.

See more of this Session: Model Based QbD In Drug Substance or Drug Product Development
See more of this Group/Topical: Topical I: Comprehensive Quality by Design in Pharmaceutical Development and Manufacture

(701g) Controllability Analysis of Protein Glycosylation In CHO Cells

Melissa M. St. Amand, Kevin Tran, Devesh Radhakrishnan, Anne S. Robinson and Babatunde A. Ogunnaike, Chemical Engineering, University of Delaware, Newark, DE

Biopharmaceuticals represent the largest growing class of therapeutics with a US market of approximately $50 billion in 2009 and an expected steady increase in future sales (1). Many biopharmaceuticals, such as the therapeutic monoclonal antibodies Herceptin and Avastin, are produced as recombinant proteins from Chinese hamster ovary (CHO) cells that are cultivated in bioreactors. As with other manufactured products, these therapeutic proteins are effective only when their product quality attributes (bioactivity, potency, purity, etc.) lie within a specific range of values. Of the many factors that affect the quality and bioactivity of these proteins, arguably the most important is glycosylation, a post-translational modification in which a carbohydrate chain, termed a glycan, is added to a protein. To function as intended *in vivo,* most therapeutic protein treatments validated for human use must have a precise distribution of glycans (i.e., specific percentages of glycans with specific sugar monomers such as galactose, sialic acid, or fucose). Unlike other cellular processes such as DNA replication and protein production, however, glycosylation has no master template, and, as a result, glycan formation and attachment are subject to variability and are often non-uniform. Consequently, regulatory agencies, such as the Food and Drug Administration (FDA) and European Medicines Agency (EMEA), are now encouraging biopharmaceutical manufacturers to develop strategies to control glycosylation online during production. However, online glycosylation control, which will involve active monitoring and effective control of key process variables that are critical to product quality, is yet to be implemented in the biopharmaceutical industry for a variety of reasons, mostly attributable to the complexity of these bioprocesses, the non-availability of on-line measurements, and the lack of comprehensive control paradigms tailor-made for such processes.

Our goal is to develop—and validate experimentally—such a comprehensive strategy for effective online, real-time control of glycosylation. To achieve this goal, however, it is imperative first to assess the controllability of the process of glycosylation. Specifically, with x defined as the vector of the percentages of each glycan form present in the overall glycan pool from a batch of therapeutic protein, we must determine if protein glycosylation can be directed from any initial state $x(0)=x_0$ to any arbitrarily specified desired final state x_f, in finite time, via admissible manipulations of available process variables and operating conditions. Unfortunately, even though there is sufficient mechanistic knowledge to enable the development of a first-principles process model of glycosylation, by the very nature of the glycosylation process, with its convoluted reaction schemes and complex network architecture of approximately 23,000 reactions and almost 8,000 glycoforms, it is entirely *impossible* to carry out standard theoretical closed-form, state-space controllability analysis with the resulting mathematical model. While numerical analysis is always possible *in principle*, the intrinsic nonlinearity and sheer complexity of the mathematical model demands that even model simulations must be carried out systematically and judiciously if the required system controllability information is to be extracted efficiently from the simulation results.

In this presentation, we discuss a simulation-based procedure for assessing the controllability of glycosylation predicated upon employing statistical design of experiments to vary glycosylation enzyme and sugar nucleotide concentrations (model inputs) strategically. The resulting glycosylation responses (model outputs) are analyzed via ANOVA to identify the enzymes and sugar nucleotide donors that exert the most significant effects on the predominant glycans found in recombinant proteins produced with CHO cells. The statistical analysis results are subsequently used to assess the degree to which glycosylation is controllable. In addition to providing useful insight into the extent to which online glycosylation control is achievable in practice in an actual experimental process, we also discuss how the procedure provides a framework for developing directed approaches for the genetic engineering of cell lines to produce protein with a specific desired glycan structure.

[1] Aggarwal, S., What's fueling the biotech engine-2009-2010. Nat Biotechnol, 2010. 28(11): p. 1165-71.

See more of this Session: Model Based QbD In Drug Substance or Drug Product Development
See more of this Group/Topical: Topical I: Comprehensive Quality by Design in Pharmaceutical Development and Manufacture

(752a) Quality by Design for DPIs -A Regulatory Perspective

Prasad Peri, DHHS/FDA/Cder/OPS/ONDQA/Dndqa I, FDA, Silver Spring, MD

Inhalation Powders have become an important dosage form in developing products for delivery to the lungs. Since the approval of the first DPI in the US in the late 1990's there has been a rapid surge in the number of drug development programs involving Dry Powder Inhalers. The draft MDI/DPI guidance, developed more than 10 years ago, discuss approaches for CMC information to be submitted in an application. An update to this guidance is being developed and the proposed regulatory recommendations for DPIs will suggest a Quality by Design (QbD) approach whereby a science and risk based understanding of the device design, its effect on the performance, and the patient interaction is sought. Specifically, the impact of interaction of the critical quality attributes of the actives, excipients, device components, on the Quality Target Product Profile (QTPP) of the finished product (e.g. delivered dose, aerodynamic performance of the drug product) is an example of enhanced development information that can be included in applications. These recommendations are in line with the Agency's continued efforts to promote the implementation of QbD principles in drug development.

See more of this Session: QbD In Dry Powder Inhaler Development : Multifaceted Powder Characterization for Product Development
See more of this Group/Topical: Topical I: Comprehensive Quality by Design in Pharmaceutical Development and Manufacture

(752b) QbD In Dry Powder Inhaler Development

David Morton, Drug Delivery, Disposition and Dynamics, Monash Institute of Pharmaceutical Sciences, Monash University, Melbourne, Melbourne, Australia

Summary

Developing products comprising un-bound fine cohesive powders throws up significant challenges. Dry powder inhalers present a prime example. Why? Because the behaviour of fine cohesive particle behaviour is notoriously complex, and notably more so than any single phase solid, liquid or gaseous material (1): where powders can exhibit the behaviour of solids, liquids or gases- and sometimes all three. Furthermore, the complexity is compounded by the need to match the nature of such a complex powder, with a small dispersion device with aerodynamically complex air flow channels intended to create some form of dispersion energy, as well as the reliance on bringing the function together by the complex and variable inspiration pattern of the patient as they inconsistently inhale through these air channels to actuate the system (2).

A few have ventured to computationally model and even develop measurement or theoretically based strategies to try to predict performance of a powder within a given device. It is proposed that none can claim success, in a way that captures effective predictive capability. Product development in this area consequently remains largely based on empirical controls of the formulation, and consequently QbD implementation in its true sense remains out of reach. Successful current development is therefore largely based on parametric control of both input materials and processes, but where the influence of parameters are not fully known, they are selected without a fundamental scientific basis, and do not yield a true design space, nor any predictive capacity. Furthermore, it is also proposed here that outcomes of the measurement of basic surface properties of real (as opposed to model and ideal) materials is either too simplistic, based on unrealistic "isolated" measurements or even beyond current scope of practical technology- for example, single particle adhesion force-contacts, electrostatic charge measurement, surface rugosity, and probing molecular surface structure cannot be either usefully or effectively measured for real pharmaceutical powders. Hence the subtitle of this paper- is a provocative challenge that past research has suffered by too great a focus much on the measurement, rather than as a problem-centred approach.

In this paper, a number of case studies will be used to illustrate apparent contradictions and failures to develop even a simple predictive capability from such measurements.

However, we have on the market highly successful and effective DPI products, and progress in understanding is ongoing. Some semi-empirical approaches have also had qualified success, such as the Cohesive-Adhesive Balance approach (CAB), despite its flaws (3). This paper will then focus on some more recent problem-driven and bulk scale approaches to make more pragmatic and hopefully effective powder property measurements and models.

References

1. Igwe GJI, 1991. "Powder technology and multiphase systems: Gas permeametry and surface area measurement". New York: Ellis Horwood Ltd, Preface.

2. Morton, DAV, Staniforth, JN, 2005. "The Challenge of the New: Device-Formulation Matching in Dry Powder Inhaler Systems." Pharm. Manuf. Packing Sourcer, Spring, pp. 80–83.

3. Begat P, Morton DAV, Staniforth JN, Price R, 2004. "The cohesive-adhesive balances in dry powder inhaler formulations. I: Direct quantification by atomic force microscopy." Pharm Res 21:1591–1597.

See more of this Session: QbD In Dry Powder Inhaler Development : Multifaceted Powder Characterization for Product Development
See more of this Group/Topical: Topical I: Comprehensive Quality by Design in Pharmaceutical Development and Manufacture

(752c) The Influence of Crystals, Particles and Powders On the Performance of Dry Powder Inhalers

Jag Shur, School of Pharmacy & Pharmacology,, University of Bath, Bath, Avon, BA2 7AY, United Kingdom and Robert Price, University of Bath, Bath, Avon, BA2 7AY, United Kingdom

Summary

Dry powder inhalers (DPIs) are an attractive means of delivering drugs to the lung. The fluidization, entrainment and de-aggregation of a DPI formulation is a complex relationship between the patient's inspiratory effort, the air flow field generated within a device and the cohesive properties of the powder bed.

The inter-relationship between factors including particle size, particle shape, distribution of interparticulate and mechanical forces govern the performance and efficacy of DPI drug delivery systems. However, quantitative measurements of these properties and their under-lying relationships remain difficult to discern.

This presentation will focus on some key developments in understanding the relationships between the governing properties of DPIs and how these properties can be harnessed to develop designer formulation systems for DPIs.

INTRODUCTION

The ability to modify the flowability, deaggregation and dispersion of particles in a well-defined particle size range is a key requirement for the delivery of therapies to the lungs using dry powder inhalers (DPIs). The de-aggregation and dispersion of particles from DPIs is controlled by interfacial forces, and therefore governs the performance and efficacy of medicaments delivered to the lung by DPIs.

Several physicochemical factors can directly influence these properties, amongst which the mesoscopic surface structure of particulates plays a dominant role. However, quantitative measurements of these effects are difficult to discern, particularly as the size of particles approaches the micrometer dimensions. However, quantitative colloid-probe atomic force microscopy (AFM) measurements of the spatial distribution of the interactive forces, over and upon contact with the substrate surface have allowed changes in solid-state properties to be directly related to particle adhesion. Furthermore, this approach has enabled the ability to relate particle adhesion to the performance of DPI drug delivery systems.

Another critical component of delivery of dry powders to the lungs is related to the powder fluidisation process within the delivery device. The process of fluidization is related to many factors such as particle size, particle shape, bulk density and powder flowability. The influence of each of the components on the fluidization process is complex, however the development of a fluidisation energy measurement has provided a significant insight into specific relationship between powder flow, cohesive properties of the powder and DPI performance.

The presentation will specifically focus on the current limitations in the processing of sub-10mm particulates for DPIs, highlighting the direct influence of particle shape, surface morphology and interfacial properties on particle interactions. In addition, key developments in particle engineering techniques in modifying the particle architecture and surface geometry will be explored. The presentation will also focus on how rheometric properties of granular materials may be a critical tool in the development of powder-based formulations for inhaled drug delivery.

See more of this Session: QbD In Dry Powder Inhaler Development : Multifaceted Powder Characterization for Product Development
See more of this Group/Topical: Topical I: Comprehensive Quality by Design in Pharmaceutical Development and Manufacture

(752d) A Dry Powder Inhaler, a User Interface or a Formulation Interface? Designing for Quality Amidst Competing Requirements

James Tibbatts, Concept Flow Ltd, Cambridge, United Kingdom

A dry powder inhaler is the physical interface that sits between the patient and the powder formulation. Faced with the device, the patient must go through a number of steps to prime and prepare the dose, before inhaling on the device. During the inhalation, the patient's inspiratory energy is used to break up the powder and to deliver the powder to the lungs. Clearly, both the inner workings of the inhaler and the manner in which that inhaler is used have a significant and direct influence on the clinical efficacy of the product. However, this raises two competing design requirements which are often dealt with as separate design activities.

Laboratory testing shows that, when correctly used, the performance variation between different inhalers, and at different flow rates through the same inhaler, is significant. When user error and compliance issues are considered on top of that inherent variability, the need for quality design is abundantly clear. However, true QbD to mitigate this variability is perhaps an unrealistic proposition. Despite this, there remain a wide range of poorly understood factors which are central to the DPI performance where improved understanding and characterisation can bring us a step closer to QbD.

The physics that determines performance is multifaceted and is influenced by turbulence, shear, impaction, entrainment, deaggregation and deposition. Compounding this complexity is a lack of available methods for direct characterisation of these factors. As a result, the DPI airpath is often developed as a consequence of, and not in concert with, usability and mechanical design requirements. This approach effectively prioritises the aspects of the design that can be more readily quantified over and above the aspects of the design that are poorly understood. Instead of improving quality, this design paradigm simply hides the unknowns.

This paper will focus the characterisation of the airflow within the device and on the characterisation of the powder behaviour as a result of that airflow. The analysis will focus on new approaches for both in-silico and in-vitro data analysis that can help to unpick the causal relationships that determine DPI performance. The data also highlights the detrimental influence that competing design requirements of usability and performance can have on the performance and consistency of the product. The presented methods provide a set of tools that can be used to improve the quality of DPI design, and that can be used provide a central focus for disparate design teams.

(754a) Application of a Joint Risk Assessment Platform to Identify Suitable Drug Substance and Drug Product Process Options

Joshua D. Engstrom[1], Jonathan Brown[1], Olav Lyngberg[2], Lindsay Hobson[3] and San Kiang[1], (1)Drug Product Science and Technology, Bristol-Myers Squibb Company, New Brunswick, NJ, (2)Early Phase Chemical Development, Bristol-Myers Squibb Company, New Brunswick, NJ, (3)Late Phase Chemical Development, Bristol-Myers Squibb Company, New Brunswick, NJ

The final selection of a robust drug substance process in combination with a drug product process can be extremely challenging, especially in cases in which the crystal form and powder properties of the active pharmaceutical ingredient (API) directly impact the development of a robust and economic drug product process. For such cases, it is necessary to assess and manage the overall risk of a given drug substance (API) + drug product combination instead of analyzing drug substance and drug product risks independently. In this presentation a case study is described in which a joint risk assessment platform was organized to rank combinations of various API crystal form inputs with drug product processes. The joint risk assessment was executed in the following phases: planning, data collection for each API + drug product option, data review and scoring, and final API + drug product ranking and review. Data collected from lab feasibility studies on each option were utilized to help more accurately score the perceived processing risks in each unit operation. The risks were scored using a failure modes and effects analysis (FMEA) quality risk assessment tool. Through the joint risk assessment approach, appropriate API + drug product options were identified to help the drug substance and drug product teams effectively direct resources toward processes that have the highest chance of successfully delivering a commercially manufactured drug product.

See more of this Session: Scale-up of Pharmaceutical Manufacturing Processes: Toward a QbD Approach
See more of this Group/Topical: Topical I: Comprehensive Quality by Design in Pharmaceutical Development and Manufacture

(754b) Improved Process Understanding Through Implementation of QbD Methodologies

Shane T. Grosser[1], Aaron Moment[2], Angela Spartalis[2], George Zhou[3] and Paul Fernandez[2], (1) Chemical Process Development and Commercialization, Merck & Co., Inc., Rahway, NJ, (2)Chemical Process Development and Commerciliazation, Merck & Co., Inc., Rahway, NJ, (3)Analytical Development and Commercialization and Development - Specialty, Merck & Co., Inc., Rahway, NJ

Purpose: The work presented here was conducted to develop a process for the pure step of an ongoing developmental program that simultaneously provides operational flexibility and sufficient process robustness inline with QbD filing strategies. Particular challenges of this process are the strong dependence of final API PSD on seed temperature and the significant impact that solvent composition has on that temperature. Additional steps have been investigated to limit the variability of residual solvents going into the pure step and therefore simplify the seed point determination. Unfortunately these steps have both yield and time-cycle implications and could unnecessarily impact productivity for batches when residual solvent levels may already be low. As a result, the project team has decided to develop a mechanism by which the optimal seed temperature is determined in real-time based upon batch composition assays, therefore enabling increased flexibility in solvent composition without comprising batch quality.

Methods: This work will demonstrate the application of risk assessment strategies for targeted experimental planning and implementation of statistically designed experiments (DoE) for investigation of interacting process parameters. We will also discuss the use of pilot-scale confirmatory runs of DoE results and how large scale experimentation was used to drive further DoE investigation.

We will present the use of PAT technologies for the rapid collection of solubility data to aid in mapping API solubility as a function of solvent composition for a 4-solvent system.

Lastly, we will discuss the use of thermodynamic and empirical modeling tools for the translation of the acquired solubility map into operational parameter ranges and assay specifications to be implemented in the full-scale demonstration and subsequent filing.

Results: Guided by a formal risk assessment, significant DoE experimentation was performed aimed at understanding the effect that 9 operational parameters have upon the API PSD, purity and form. None of the investigated parameters showed evidence of a purity or form impact while PSD was primarily impacted by seed temperature. Although the PSD was not directly influenced by solvent composition considerations, it was readily observable that solvent composition had a profound impact on the optimal seed temperature so by extension, needed to be considered to ensure API PSD within specification.

To facilitate understanding of the relationship between solvent composition and optimal seed temperature, the crystallization platform was employed to construct an API solubility map across the 4-dimensional solvent composition space. With the use of the crystallization platform, the collection of solubility as a function of temperature was automated for each solvent composition. The use of PAT technologies helped to attain the desired data while avoiding the complication of offline sampling of a solution susceptible of crystallization upon cooling, thus greatly reduced the time and effort required to construct this solubility map.

With the relationship between solvent composition and API solubility measured at discrete solvent compositions, the next step was to convert that map into a model that allowed investigation of the continuum of solvent compositions studied. Two strategies were investigated. In the first strategy, a thermodynamic model was used to predict the API solubility. The solubility model employed was based upon the Unifac activity coefficient model and the data obtained through the PAT investigation was used to regress the binary interaction parameters between the API and the solvent subgroups. Secondly, an empirical (polynomial with three variables) model was also developed and found to offer only slightly less accurate predictions of API solubility. Due to the expected ease of implementation on the factory floor and within the filing, the decision was made to utilize the polynomial model long term. With a firm understanding of the relationship between composition and API solubility and an eye to expected process variabilities, the allowable seed temperature range could be determined and ranges on the allowable solvent levels could be specified.

Conclusions: Through efficient application of QbD methodologies, the extended development team has established a process that enables significant operational flexibility and process robustness for an otherwise challenging seed point determination. Although the seed temperature is critical in ensuring the final API PSD, a procedure has been put in place to ensure an appropriate seed temperature while simultaneously leveraging the flexibility in solvent composition afforded through QbD.

See more of this Session: Scale-up of Pharmaceutical Manufacturing Processes: Toward a QbD Approach
See more of this Group/Topical: Topical I: Comprehensive Quality by Design in Pharmaceutical Development and Manufacture

(754c) Harnessing Computational Fluid Dynamics (CFD) to Guide Reliable Scale-up of Pharmaceutical Processes

Ann M. Czyzewski, Process Engineering, Abbott Laboratories, North Chicago, IL, Hsien-Hsin Tung, Global Pharmaceutical Research & Development, Abbott Laboratories, North Chicago, IL, Shailendra V. Bordawekar, Process R & D, Global Pharmaceutical Research and Development, Abbott Laboratories, North Chicago, IL and **Nandkishor K. Nere**, Process R&D, Global Pharmaceutical Reseach and Development, Abbott Laboratories, North Chicago, IL

Successful scale-up of pharmaceutical batch and fed-batch processes in stirred vessels is largely a function of mixing that is governed by underlying hydrodynamics. This in turn is a strong function of various geometric and operational parameters, namely reactor diameter, fill level, impeller (type, number, size, location, and speed), the degree of baffling, feed location and addition rate along with physicochemical properties of the fluid and processing conditions (temperature and pressure). Reactor and impeller geometric configurations as well as vessel internals in laboratory reactors are significantly different than those at the pilot and commercial scales. Hence, it is crucial to understand the differences in the hydrodynamics across scales for accurate process scale-up. To this end, computational fluid dynamics (CFD) provides an efficient and reliable means for characterizing fluid flow.

We present general cases where the information obtained from CFD simulations carried out for a variety of configurations across scales can be effectively used to develop process understanding and guide reliable scale-up of batch and fed-batch processes.

See more of this Session: Scale-up of Pharmaceutical Manufacturing Processes: Toward a QbD Approach
See more of this Group/Topical: Topical I: Comprehensive Quality by Design in Pharmaceutical Development and Manufacture

(754d) Model-Guided Quality by Design: A Case Study On a Pharmaceutical Final Intermediate

Nathaniel D. Kopp[1], Zhongmin Xu[1], James S. Bergum[2], Thomas L. Laporte[1], Jose E. Tabora[1] and Lori A. Spangler[1], (1)Chemical Development, Bristol-Myers Squibb Company, New Brunswick, NJ, (2) Global Biometric Sciences, Bristol-Myers Squibb Company, New Brunswick, NJ

This case study outlines an application of Quality by Design (QbD) principles to the development of the final intermediate manufacturing process. The final intermediate process consists of three steps: (1) deprotection, (2) alkylation and (3) recrystallization. This study will focus on the alkylation step which affords two key impurities and requires control of three process parameters: temperature, solvent ratio and reagent equivalents to produce high quality material.

Through a multi-staged experimental plan, a statistical model was developed and refined to guide the selection and verification of the manufacturing design space. Early Design of Experiments (DoE) laboratory work established the dependence of the quality attributes on specific process parameters which were investigated further to develop an initial statistical model This model was further refined through incorporation of additional laboratory and production-scale data to provide guidance for the determination of the optimal manufacturing design space. The 95% prediction intervals of the model were utilized as an initial guide for the design space. The design space was verified in the laboratory and at pilot scale. This presentation will discuss the implementation of QbD principles to maximize the manufacturing process robustness for the alkylation step through experimental design, statistical model development and design space verification.

See more of this Session: Scale-up of Pharmaceutical Manufacturing Processes: Toward a QbD Approach
See more of this Group/Topical: Topical I: Comprehensive Quality by Design in Pharmaceutical Development and Manufacture

(754e) Incorporating Computational Fluid Dynamics Methods for Quality by Design Framework for Exploring Design Space of a Fermentor

Sravan Kumar Nallamothu, ANSYS Inc, Pune, India and **R. Muralikrishnan**, Ansys, ANSYS Inc, Pune, India

In the Quality by Design (QbD) framework, there is a significant emphasis on the robust characterization of manufacturing processes and prospectively identifying the engineering design space that ensures product quality. Although the QbD approach is expected to bring about significant long-term benefits, there are concerns regarding the increased amount of characterization work that will be required upfront, which can be very resource intensive. In cases where the number of input variables is large, performing a design of experiments (DOE) with statistical significance may be impractical.

Computational fluid dynamics (CFD) provides a first principles approach to gain insight into the hydrodynamics and mass transfer. The scope of traditional CFD methods can be extended to explore the engineering design space of the mixing process by coupling to statistical tools. In this talk, the authors summarize the results of a multivariate study of a prototypical fermentation process. And while a direct link to product quality cannot be provided by CFD alone, the authors hope to illustrate how CFD coupled with statistical methods can lower the risk of this unit operation having an adverse impact on product quality.

See more of this Session: Scale-up of Pharmaceutical Manufacturing Processes: Toward a QbD Approach
See more of this Group/Topical: Topical I: Comprehensive Quality by Design in Pharmaceutical Development and Manufacture

(754f) Scaling up Strategy for Continuous Powder Mixing Process

Yijie Gao, Chem and BioChem Engineering, Rutgers, the State University of New Jersey, Piscataway, NJ, Fernando. J Muzzio, Chemical and Biochemical Engineering, Rutgers University, Piscataway, NJ and Marianthi G. Ierapetritou, Department of Chemical and Biochemical Engineering, Rutgers, The State University of New Jersey, Piscataway, NJ

Scaling upstrategy for continuous powder mixing process

Yijie Gao, Fernando Muzzio andMarianthi G. Ierapetritou

Dept. of Chemical and Biochemical Engineering, RutgersUniversity, Piscataway, NJ 08854

Continuouspowder mixing has attracted a lot of interest among pharmaceutical industry. Basedon the principle of similarity proposed by Johnstoneand Thring [1],many works have been done recently that target the scale-up of continuouspowder mixing. In this work, a quantitative scale-up strategy is described thatallows the transition from lab to industrial scale. By using the periodicsection modeling developed in our previous study [2], scale-up of the cross-sectional mixing and scale-up of theaxial mixing are separately considered. To capture the cross-sectional mixingcomponent, the conception of variance spectrum is used. The Lomb-Scargle Periodogram techniquewidely applied in astronomical studies has been introduced to estimate thevariance spectrum of unevenly distributed samples in powder mixing process. Theeffect of constant sample size on mixing performance of different scales is thuselucidated. To capture the axial mixing component, the residence timedistribution is used to characterize mixing of different scales.

Two casestudies are presented to illustrate the applicability of the developedstrategy. For non-segregating mixing case, similar scale-up characteristics areobserved for both cross-sectional and axial mixing. While satisfying the principleof similarity leads to similar decay contours of variance spectrum, a samplingsize much smaller than scales of both mixers should be used to achieve similar measurementsof mixing performance. A 2.5 power increase of flux rate is generallyobtainable in the scale-up of non-segregating materials. However, in processeswhere cohesive materials are used, segregation occurs due to a tendency ofagglomeration, and a larger increase of flow rate is observed. This resultsfrom the effect of cohesion on axial motion especially when the scale of mixingprocess is small. Our method can be used to provide specific guidance of scale-upof different powder mixing cases.

Reference

1. JohnstoneRW, Thring MW, *Pilot plants, models andscale-up methods in chemical engineering.*

1957, MacGraw-Hill: New York.

2. GaoY, Ierapetritou M, Muzzio F. Periodic section modeling of convective continuouspowder mixing processes. *AIChE Journal*.2011; . In Press: DOI: 10.1002/aic.12563.

See more of this Session: Scale-up of Pharmaceutical Manufacturing Processes: Toward a QbD Approach
See more of this Group/Topical: Topical I: Comprehensive Quality by Design in Pharmaceutical Development and Manufacture

(754g) Modified Application of Johanson's Model towards Roller Compaction

Jasmine M. Rowe and John R. Crison, Drug Product Science & Technology, Bristol-Myers-Squibb, New Brunswick, NJ

ModifiedApplication of Johanson's Model Towards Roller Compaction

Jasmine Rowe, JohnCrison

Rollercompaction (RC) is a continuous dry granulation process commonly used in thepharmaceutical industry to improve powder flow and blend uniformity via powderdensification. RC has gained substantial attention over the last couple ofdecades because neither moisture nor heat is required for granulation (as isthe case for wet granulation and hot melt extrusion, respectively). Theprocess by which densification is achieved is through powder being fed betweentwo rolls that are rotating in opposite directions of each other. Thiscounter-rotating motion of the rolls and friction between the powder and theroll surface draw the powder through the rolls and maximum compaction(densification) occurs when the separation between the rolls is at itsminimum. The resultant compact, or briquette, may then be milled to achievethe desired granule size.

While the RC mechanical process appears tobe simple, efforts to quantitatively model this process have proven challengingbecause of complex material behavior in the feeding and compaction zones. Consequently, a trial-and-error approach has typically been adopted towards RCprocess design, which is both time- and resource-intensive. In response to theneed for a more efficient approach towards RC processing, Johanson developedone of the most widely-used models for RC simulation, a powder mechanics modelbased on Jenike's steady state flow theory that predicts the physicalproperties of the final compact based on material and process parameters [Johanson, J. R. (1965).A rolling theory for granular solids. ASME, Journal of Applied Mechanics SeriesE, 32(4), 842 848]. However, practical application of Johanson's model to experimental work hasbeen limited because one of the model input parameters, the pressure beingapplied to the feed powder as it enters the rolls (P_0), is difficultto obtain experimentally without sophisticated pressure sensor-instrumentedequipment.

In this work, an alternative approach hasbeen established, which expands upon Johanson's model to enable itsincorporation into a daily work flow. This new method requires only standard,routinely-measured, process parameters as inputs and is capable of solving for P_0a priori to extensive experimentation. An excellent correlation between simulatedand experimental results has been achieved for both placebo and active blends. Additionally, a dimensionless relationship between key process input parametersand final compact properties has been elucidated, which provides valuableinformation for efficient scale-up procedures.

AUTHOR INDEX

AUTHOR INDEX